ERRATA   All references to page 58 should read page 144
On page 65 (see page 00) should be (see page 131)

Windmills

of

Anglesey

"Old Mill at Ty Croes"
(*Melin y bont, Bryn Du, Anglesey*)

Wood engraving from The Graphic
C.1870
From a watercolour by J. Orrock

Barry Guise and George Lees

Windmills
of
Anglesey

Illustrated by Ron Pyatt

attic books

WALES

INTRO

Published by Attic Books, The Folly, Rhosgoch, Painscastle, Builth Wells
Powys LD2 3JY
   First published 1992
   Copyright © Barry Guise and George Lees, 1992
British Library Cataloguing-in-Publication Date
   Guise, Barry
   Windmills of Anglesey.
   I. Title II. Lees, George
   621.42921
   ISBN 0-948083-16-6

Typeset in Univers by Proteus Micro Applications, Avon
Printed in England by J. W. Arrowsmith Ltd, Bristol

For Lesley and Mair

Contents

Acknowledgements

The authors wish to thank the following institutions for their help and cooperation during the preparation of this book:

Anglesey Borough Council
Gwynedd Archives Service (Llangefni and Caernarvon)
Gwynedd Family History Society
Gwynedd Library Service (Llangefni and Holyhead)
National Library of Wales, Aberystwyth
U.C.N.W. (Bangor) Archives
Welsh Folk Museum, St Fagans, Cardiff
Welsh Mills Society

We are also indebted to the many individuals who have shared their knowledge of Anglesey's windmills with us. Particular thanks are due to:

Vera Bradford
Andrew Davidson
Tom Davies
Joan and Nigel Dennis
Sean Hagerty
Evan Wyn Hughes
W.H. James
Connie Jones
Gwyneth Jones (daughter of the late Alan Ross)
Tony Parkinson
John Owen Parry
Tomos Roberts
Revd. J. Rice Rowlands
Frank Stead
M.J.A. Swallow
Mary Tucker (widow of Professor G. Tucker)
Rex Wailes (now sadly deceased)
S.F. Wallace
Phil Williams

Foreword

House of Lords

I have heard some casual observers say, after travelling across the island, that they found the landscape of Anglesey rather boring especially compared with the mainland. It may seem flat and uninteresting to the shortsighted traveller on his way to Holyhead and the cross channel boat to Dun Laoghaire, but I still recall the first sentence uttered by our geography teacher in the Holyhead County School: ''The island of Anglesey is not flat, it is an undulating plain''. In 1907 David Lloyd George, then Member of Parliament for the neighbouring Caernarfon Boroughs, presided over the Anglesey Eisteddford in Llanrhyddlad and referred to Anglesey as ''a splendid platform from which to view the glories of Caernarvonshire''. His audience, which included his parliamentary colleague Ellis Jones Griffith the Member for Anglesey, knew that Lloyd George was pulling their legs and responded when he spoke later saying that ''the mountains of Caernarfon were on tiptoe enthralled by the mysterious charms of Anglesey''.

One of these charms from the eighteenth century onward has been the windmills for which the island became famous. They were a picturesque feature of the landscape as well as an essential source of food for a growing population. There were also numerous watermills on the island but these were less reliable because an adequate supply of water to work them could not always be guaranteed. But as those of us who live here know too well another natural element, namely the wind, is never in short supply.

In little more than a century the number of working wind and watermills has fallen from a hundred to two, Llynon windmill and Howell watermill, both in Llanddeusant. We are greatly indebted to the Anglesey Borough Council for their enterprise in purchasing Llynon Mill in 1978 when it was in a state of advanced disrepair and restoring it until it is today a windmill in full working order which attracts a large number of visitors every year.

Barry Guise and George Lees, both members of the Welsh Mills Society, have made an extensive study of the windmills of Anglesey over many years and this fascinating book is the fruit of their labours. As a boy I have a vivid recollection of the sails of the Stanley Mill in Trearddur flailing away in the south-west wind. It ceased to work in 1938 when it was almost destroyed by November gales. It was the last windmill to work in the whole of Wales and, I think, in a good part of England as well. The authors have brought the old mills back to life and produced a book of significant local interest which deserves to be widely read.

CLEDWYN OF PENRHOS

Introduction

'There is nothing of note to be seen in the Isle of
Anglesea but the town, and the castle of
Beaumaris.'

DANIEL DEFOE'S summary dismissal of Anglesey, which appeared in his book
A Tour through the Whole Island of Great Britain published in 1724, is frequently echoed by more modern travellers as they hurry across the island intent only on reaching Holyhead in time to catch the ferry to Ireland. There is,
it has to be admitted, a certain justification for this opinion. Thomas Telford's
cross island link — the present A5 — was designed to be efficient rather than
scenic and consequently does not traverse the most attractive parts of Anglesey,
nor are the grey villages that were drawn to its edge the most prepossessing
of the island's settlements.

But to judge Anglesey on the views obtained from a car speeding along the
busy A5 would not be fair. For those willing to turn off Telford's highway a
different Anglesey awaits discovery, one of whitewashed cottages and peaceful
farms, ancient burial mounds and standing stones, isolated churches and country
houses, not to mention unrivalled coastal scenery of rugged cliffs, secluded
bays and sheltered creeks.

Defoe had crossed the Menai Strait from Caernarvon on his way to Holyhead
'for no purpose but to have another view of Ireland'. Travelling thence to
Beaumaris he felt obliged to temper his original opinion of Anglesey, conceding
that 'it is a much pleasanter country, than any part of N. Wales, that we have
yet seen; and particularly is very fruitful for corn and cattle'.

Nearly a hundred years later Nicholson, author of *The Cambrian Traveller*,
also looked beneath the island's superficial dullness, commenting 'the country assumes a very dreary aspect suggesting the idea of sterility; yet this is not
the real state of the case, for the soil is generally very good, and, under proper
management, highly productive. Corn and cattle constitute the chief products'.

Corn and cattle were the mainstays of Anglesey's agriculture and, for hun-
dreds of years, provided the foundation of its economic prosperity, earning the
island its reputation for fertility. Writing towards the end of the twelfth century
Giraldus Cambrensis described Anglesey as 'incomparably more fertile in corn
than any other part of Wales, from whence arose the British proverb, 'Môn
mam Cymry, Mona mother of Wales', and when the crops have been defec-
tive in all other parts of the country, this island, from the richness of its soil
and abundant produce, has been able to supply all Wales', — a statement which

Windmills of Anglesey

probably says more about the size of the Welsh population than the fertility of Anglesey!

Exaggerated though the claim 'Mother of Wales' might have been, the strategic importance of Anglesey's grain production was not lost on Edward I in his campaign to subdue the Welsh in the late thirteenth century. At that time Anglesey formed part of Gwynedd (one of several political states into which Wales was divided), with the Royal Court established at Aberffraw in the south-west of the island. By invading Anglesey and seizing the harvest Edward I was able to force Llywelyn ap Gruffydd to surrender, an indication of the crucial role played by the island in the economy of Gwynedd.

Whether or not the epithet 'Mother of Wales' was ever true is open to debate, but there is no doubting the fact that compared with the barren nature of mainland Wales Anglesey was, and still is, extremely fertile. Nevertheless, the reality is that even by the beginning of the eighteenth century only a very small percentage of the island was under cultivation. Many parts were unsuited to arable farming: the centre was ill-drained and marshy, much of the north and west had thin soils and rocky outcrops, while wind-blown sand covered extensive areas of the south-west.

Despite these drawbacks, Anglesey still managed to produce more grain than any other region in Wales. This can be attributed to the fact that the island possesses certain geographical advantages with regard to grain cultivation.

Essentially low-lying, Anglesey's landscape is one of gentle contours interrupted only by a few isolated hills on the periphery, such as Mynydd y Garn (555'), Mynydd Bodafon (583') and Holyhead Mountain (720').

Thanks to its maritime position Anglesey enjoys a mild, equable climate. Spring comes earlier than on the mainland, and both temperatures and sunshine hours are among the highest in Wales. Rainfall, on the other hand, is low, the moisture-laden winds tending to pass over the island before depositing their rain on the mountains of Snowdonia.

Contrary to popular belief, Anglesey's soils are not of exceptional quality, although they are moderately fertile and amenable to improvement. The best soils occur in the south-east of the island and it is here that the highest yields have traditionally been obtained.

As Anglesey's population increased during the eighteenth and early nineteenth centuries so the acreage devoted to grain expanded. Windmills were erected to provide the additional grinding capacity required, existing watermills being unable to cope with the rising demand. Such was the amount of grain being produced that Nicholson remarked in 1813 'In favourable seasons large quantities of barley and oats are exported, either to Liverpool or across the Menai'.

Introduction

On Anglesey barley and oats were grown in preference to wheat, being the basis for the traditional bread, cakes and porridge which constituted the staple diet of the island's poor.

The exposed nature of Anglesey makes it ideally suited to windmills. There are few calm days and the open countryside offers little resistance to the strong winds which sweep in from the Irish Sea. Although a small number of windmills had been erected prior to the eighteenth century, it was between the years 1735 and 1835 that windmill construction flourished on Anglesey. At the end of this period over forty windmills were at work on the island.

However, this picture was soon to change. From around 1850 cheap foreign grain began entering the country in increasing amounts, causing the prices paid for home produced grain to slump. Falling incomes led a growing number of Anglesey farmers to convert their arable land to pasture and to concentrate more and more on animal husbandry. The contraction in grain cultivation was accompanied by a corresponding decline in the fortunes of Anglesey's windmills. With less grain to mill and increasing competition from other, more efficient sources of power the days of the island's windmills were numbered. By the end of the First World War only a handful were still in operation and these were to close before many more years had elapsed. Now just one working windmill survives, the recently restored Melin Llynon at Llanddeusant, a solitary reminder of the time when Anglesey was known as the 'granary of Wales'.

Considering the important role played by the windmill in Anglesey's economic history, surprisingly little has been written on the subject. This book, while not claiming to be comprehensive (the authors are only too aware of the gaps which exist), attempts to shed some light on a neglected aspect of the island's heritage. Emphasis has been placed on the tower windmills of the eighteenth and nineteenth centuries as it is their remains which can still be seen in the landscape today. Technical terms have been kept to a minimum and those readers wishing to discover more about the history and technology of windmills in general are referred to the publications listed in the bibliography. Owing to the fragmentary nature of some of the source material a certain amount of inference and conjecture has proved unavoidable, and the authors accept responsibility for any contentious facts or opinions.

Chapter One

Origins, Development and Decline

LITTLE IS KNOWN about Anglesey's mills before the fourteenth century since there is no record comparable with England's Domesday Survey. The island's relative remoteness from England delayed the first Norman incursion until 1088, two years after the appearance of Domesday Book. The few hand grindstones, or querns, that have been unearthed belong to the pre-medieval period and while there is evidence that some horse-driven stones were in use it is probably safe to assume that, as in England, the majority of Anglesey's mills in the eleventh century were water-powered.

Following the Edwardian conquest of North Wales in the late thirteenth century royal officials carried out detailed investigations to ascertain the extent and value of the lands now in the ownership of the Prince of Wales. Information on Anglesey was included in a survey known as the **Extent of Anglesey and Caernarvonshire** drawn up by John de Delves in 1352. It revealed that the island contained over sixty mills of which seventeen were royal mills. Although less numerous than private mills, the royal mills received a more detailed coverage because they were the ones to which the Prince's tenants were obliged to take their corn to be ground, paying a fixed toll (*multure*) for the privilege. Thus his tenants at Aberffraw 'owed suit at the Prince's mill at Aberffraw . . . and grinded their corn to the thirtieth part', a toll which compared favourably with the sixteenth part common in England.

There was no statutory obligation for tenants to comply with this practice but tradition and custom meant they usually had little option. A mill was a valuable source of revenue and the Prince was unwilling to tolerate competition. Any recalcitrant tenant could be subject to a fine (*amercement*) or be punished in some other way. For his part the Prince had to ensure that the mill was adequate for the needs of his tenants and be properly maintained, otherwise they were at liberty to go elsewhere. This proviso only led to the tenants themselves being made responsible for the upkeep of the mill. For example, those in the township of Bodynwy near Llanddona 'owe suit thence to the lord prince's mill of Cefn-coch. And they carry timber and grindstones for the said mill within the county of Anglesey at their own expense. And they make the watercourse and ditch of this mill'.

Tenants, nor surprisingly, came to resent these demands and evasion was common as courts became more reluctant to prosecute offenders. The problems of enforcing mill suit led to its gradual demise, although as early as the fourteenth century in some areas it had been commuted to a money payment in lieu of toll or service.

Not all of Anglesey's mills were in royal hands. Some belonged to prominent freemen, while others were owned collectively by a family group (*gwely*) as was the case in the township of Twrgarw near Llangoed, where 'there is one free gwely called Gwely Tudor ap Madog . . . and they render nothing to the lord prince each year from it . . . and they have their own mill called Melin Tudor'.

The church also owned mills, at least until the dissolution of the monasteries when they became the property of the king. The earliest reference to a watermill on Anglesey is an ecclesiastical one, the mill at Tal-y-bont in Menai being granted to the monks of Aberconwy at the end of the twelfth century.

It may be difficult to conceive how such a relatively flat island as Anglesey could have provided sufficient water power to drive so many mills. While there are numerous rivers (such as the Afon Goch, Lligwy and Ceint in the east; Alaw, Crigyll, Cefni and Braint in the west) none is large and few flow fast enough to turn a waterwheel. Consequently mill ponds often had to be constructed to maintain a regular supply. This is confirmed by the accounts for work done on royal mills which include payments for cleaning mill ponds, repairing dams and building new sluices.

Nevertheless, the use of these mills was still restricted during very dry periods when water supplies were scarce. On Holy Island, for example, there was only enough water to supply the mill at Poth y felin. In an attempt to overcome this deficiency a mill was built at Tre'rgo (near Trearddur Bay) which utilised the ebb and flow of the tide between Holy Island and the mainland. Felin Heli's mill wheel was turned by water channelled from a dammed creek which had been filled by the rising tide. The fast tidal flows along the Menai Strait were also harnessed in a similar manner. Porthaethwy mill near Church Island contained two waterwheels, one driven by the incoming tide and one by the outgoing tide. A major drawback to this type of mill, however, was that the miller was compelled to work irregular hours, day and night, being dependent on the changing times of the tides.

Although most mills on Anglesey during the medieval period were water-powered, windmills were not entirely absent. As early as 1303 there is a record of a royal windmill being erected in the newly established settlement of Newborough. According to the Bailiff of Newborough's accounts it cost £20.12s.2½d. and began to work on 28th June. The wages of the three millers varied during the year, being highest at harvest time when the multure amounted to five bushels of barley, half a bushel of malt and six bushels of oat malt. Despite the costs and quantities of the materials used being given in the accounts there

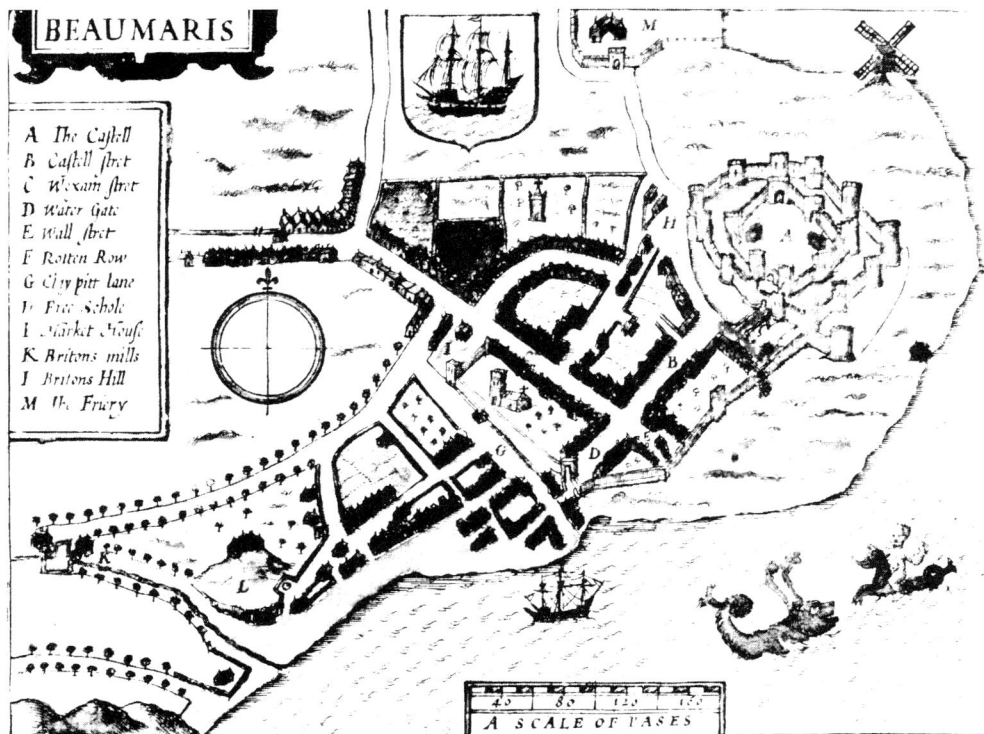

The post mill at Beaumaris.

is, unfortunately, no illustration nor description of the completed mill's appearance. The fact that the timber for its construction came by sea from Lleyn suggests that suitable wood was in short supply on Anglesey. However, this did not seem to deter others from taking advantage of the island's windswept conditions.

In 1327 Einion ab Ieuan of Beaumaris was permitted to build a windmill on the Mill Hill by the town, while an Inquiry of 1495 revealed that Rhys ap Llwelyn Hwlayn, the Sheriff of Anglesey, had erected a windmill with four sails, but where was not known. On the dissolution of the abbey of Aberconwy its property was found to include a windmill at Rhuddgaer in Menai.

Two late sixteenth century allusions to windmills occur in the Baron Hill Estate Papers. One entry notes that the tide mill at Porthaethwy was erected in 1578 and refers to land in Rossayre 'upon which a windmill was once built'; the other, from 1589, mentions 'the windmill field close to Beaumaris'. Whether this was the site of the Sheriff's mill is unclear, but the windmill is almost certainly the one depicted on John Speed's map ·of Anglesey dated 1610. The inset of Beaumaris clearly shows a wooden post mill north-east of the town next to the shore.

Post mills of that time were small, box-like structures, consisting of a wooden body which pivoted on a massive upright post supported by a trestle of cross trees and quarter bars. In order to face the wind the whole body of the mill had to be pushed round by hand using a tailpole which protruded through the ladder at the rear of the mill. The sails were simple wooden frameworks over which cloth was spread, an operation performed from ground level each sail having to be brought round to its lowest position. This, of necessity, limited the size of the sails and meant that they were capable of generating power sufficient only to drive a single pair of millstones. Such mills were easily blown over in storms. Fire was an ever present hazard and the wooden fabric of the mill·was also susceptible to attack by rot. Faced with these problems it is hardly surprising that none of Anglesey's post mills has survived.

The drawbacks of post mills meant they played only a supporting role in the rural economy and watermills continued to dominate until the early eighteenth century when the first of the tall tower windmills began to appear in the Anglesey landscape. It is probably that, as in other parts of Wales, a number of small, squat tower mills had been erected earlier but evidence of their existence remains scant. Such a mill can be seen in a 1785 painting of Parys Mountain (Plate 71), while the ruin of a straight-sided tower (now demolished) at Berw could have been another example. However, the case for the latter is dubious; a survey published in 1908, when the tower still stood, concluded 'We fail to perceive in this structure any indications of its ever having been a mill'.

Despite its reputation for corn growing Anglesey was, at the beginning of the eighteenth century, a rather backward county whose largely self-contained rural communities practised what amounted to little more than subsistence agriculture. As well as the natural constraints of climate and soils, progress was inhibited by geographical isolation, lack of resources and a stubborn reluctance to discard traditional methods.

In 1700 the island's population was estimated to number 22,800 and as the century unfolded a steady increase took place, so that by the time of the first official census in 1801 the total had risen to 33,806. Although this growth was small compared with the national average the accompanying demand for food, especially the staple crops of barley, oats and wheat, provided the stimulus for more land to be brought into cultivation. It soon became apparent that the existing watermills were inadequate to meet the needs of the growing population, a situation exacerbated by a run of dry years towards the middle of the century which resulted in corn having to be sent to the mainland for grinding.

Weather conditions during these years were recorded in the diary of an Anglesey squire named William Bulkeley. He lived at Brynddu near Llanfechell in the north of the island and, like most farmers, practised a mixed economy but with an emphasis on grain. Thus the effect of the weather on his crops was of more than passing interest. In the 1730s most of the harvests were good and surpluses were available for export. Bulkeley was able to note on 20th February 1739: 'The Corn Merchants that bought Barley for exportation gave for it, from 12s. to 13s. a Pegget [8 bushels], and three or four vessels have already laden at Cemaes and gone for Liverpool or Warrington'.

However, weather conditions deteriorated with the new decade and Bulkeley, increasingly concerned about water supplies, wrote on 18th August 1741: 'This Summer and Autumn exceeds in heat and dryness all the Summers in the memory of man, for not only all the fresh water mills are dryed thro' the country but also in a manner all the rivers and most other springs'.

The growing population, increasing acreage under grain and unreliability of water supplies provided incentives for the construction of a new generation of windmills to meet demands for additional grinding capacity. Technical advances enabled larger and more durable structures to be built capable of housing two or more pairs of grindstones. These were the tall, tapering tower mills which were to become familiar features of the Anglesey countryside. Strongly made from local stone they stood four or five storeys high, their sails carried on a revolving cap which could be turned into the wind by means of an endless chain.

The erection of such a mill was a notable event, worthy of entry in Bulkeley's diary. He recorded that on 8th September 1737: 'This day laid the foundation of the Wind Mill Tower at Allt-pen-ddu by Llanerchymedd'.

Its completion took just over a year, for according to the Baron Hill Papers 'The Windmill built upon Gallt-y-Benddu [spelling in those days was not uniform] began to grind corn on Thursday October 24th, 1738, at or near 8 o'clock'.

Sometimes Bulkeley included information about the builder of a mill, as in an entry for 24th September 1741: 'Last Saturday finishing the Wind Mill Tower of Llangoed, erected by Henry Williams of Trosmarian, being 8 yards high above the surface'.

During the last quarter of the eighteenth century the general upward trend in grain prices coupled with the protection from cheap imports afforded by the Corn Laws encouraged farmers to further expand production. Marginal land was reclaimed and enclosed, yields were raised by the application of lime and other fertilisers — and more windmills were built. Melin Llynon at Llanddeusant, completed in 1776, belongs to this period, as does Melin Maelgwyn at Bryndu, the date 1789 being inscribed over a door lintel.

Windmills of Anglesey

Date stone at Melin Maelgwyn, Bryndu.

Windmills must have been a common sight then, several being mentioned in a contemporary rhyme . . .

> *'Melin Llynon sydd yn malu,*
> *Pant-y-Gwydd sy'n ateb iddi,*
> *Cefn Coch a Melin Adda,*
> *Llanerch-y-medd sy'n malu ora.'*
> or
> *'Llynon Mill is grinding,*
> *Pant-y-Gwydd is answering,*
> *Cefn Coch and Adda Mill,*
> *Llanerch-y-medd grinds best of all.'*

Nevertheless, agricultural progress was still slow. A survey of 1795 complained that less than 10% of land in Anglesey was under cultivation and accused farmers of being 'ignorant and indolent'.

In 1793 Britain went to war with France. Fearful of being starved into submission the government urged farmers to grow more food. Unfortunately, the war years of 1793–1815 coincided with less than favourable weather conditions and the resulting poor harvests caused prices to rise to unprecedented levels. Flour, for example, which had been 3s.4d. a bushel in 1760, reached 8s. in 1794 and 10s. in 1804. Napoleon's blockade made little difference, in fact Britain even received imports from France when that country had a surplus to dispose of!

After the war new Corn Laws were introduced opening a phase of much higher protection for home producers, a move which did not unduly upset Anglesey farmers, their grain being considered inferior to that imported from the continent. Although prices fluctuated in the immediate post-war period they tended to remain relatively high on Anglesey, cushioned as it was from outside competition by its geographical isolation and the high cost of transport from other farming areas.

Melin y Gof, Trearddur Bay.

These years witnessed the culmination of windmill construction on the island, exemplified by the mills at Trearddur Bay, Llangefni, Kingsland, Cemaes and the combined wind and water mill at Bryndu near Llanfaelog. However, for some reason, little attempt was made to incorporate the recent advances in windmill technology into these new mills. Anglesey's insular millwrights shunned the fantail that automatically turned the cap into the wind, and the new patent sail with its self-regulating shutters, preferring to persevere with the traditional but labour intensive endless chain and common sail.

Windmills of Anglesey

Special mention should be made of Amlwch Port Mill. Begun towards the end of the Napoleonic War on land next to the harbour it was opened in 1816 and with seven storeys was the tallest windmill to be built on Anglesey. Amlwch had grown considerably by this time as a result of the copper mining activities on nearby Parys Mountain. It also served a prosperous agricultural hinterland and by the 1830s had developed into an important centre for the export of grain, especially to Liverpool. Anglesey's output of barley, oats and wheat had tripled since the 1795 land survey and by the 1840s the large exports of these crops from Amlwch merited the operation of five corn merchants in the town.

Windmills were then so numerous that the Reverend Robert Roberts wrote of seeing an endless succession between Bodedern and Llanerchymedd, and that fifteen could be observed from Llanddeusant alone.

But this situation was soon to change. Although the island's physical isolation had been ended by the opening of Thomas Telford's suspension bridge across the Menai Strait in 1826, the carriage of goods by road remained slow and expensive. The completion of the Chester to Holyhead Railway in 1850 was a different matter, for it brought about a dramatic reduction in transport costs to and from Anglesey.

After 1850 demand for meat products from the expanding urban centres caused livestock prices to rise markedly, encouraging farmers to convert arable land to pasture. This movement away from grain was accelerated by increasing labour costs due to shortages resulting from the migration of many low paid farm workers to the towns, enticed by the prospects of higher wages and regular hours.

By the time George Borrow's book **Wild Wales** was published in 1862 pastoral farming so dominated Anglesey's economy that when he remarked that his visit to the island had no connection with livestock an incredulous acquaintance exclaimed 'it may be so, but I can't conceive how any person, either gentle or simple, could have any business in Anglesey save that business was pigs or cattle! The advantages of animal husbandry were further enhanced by the rapid fall in grain prices which occurred from the 1870s as imports of low cost foreign grain, notably from America, began to flood the market. Without the protection of the Corn Laws, which had been repealed in 1846, prices tumbled. In the last quarter of the nineteenth century wheat prices fell by 51%, barley by 42% and oats by 37%.

The advantages of animal husbandry were further enhanced by the rapid fall in grain prices which occurred from the 1870s as imports of low cost foreign grain, notably from America, began to flood the market. Without the protection of the Corn Laws, which had been repealed in 1846, prices tumbled. In the last quarter of the nineteenth century wheat prices fell by 51%, barley by 42% and oats by 37%.

These developments heralded the end for Anglesey's windmills as viable economic concerns. In the space of a few years they were rendered obsolete by the application of steam power to the milling process and the construction at ports such as Liverpool of huge steel roller-mills to process imported grain. Such mills were capable of grinding wheat fine enough to produce flour suitable for the white bread that the public demanded. Improved transport facilities enabled the products of these mills to be carried to all areas, and the fact that it was an offence for a local baker to purchase flour 'from such towns as Liverpool to the trading detriment of the mills of Anglesey' made little difference. Gradually the island's windmills which for so long had provided an essential service to their local communities were forced to close.

Melin Ucheldre on Holy Island was one of the first casualties, driven out of business by the opening of a steam mill in Holyhead. Others soon followed, unable to supply flour of sufficient quality at competitive rates. Some millers concentrated on crushing food for farm animals, but even this task disappeared as more and more farmers invested in their own machinery. New Government regulations controlling the production of flour during the First World War only worsened the situation.

As the number of millwrights diminished (few apprentices wishing to learn a trade without prospects) windmills became increasingly difficult to maintain and costly to repair. Such was the case with Melin Llynon, damaged by a storm in 1918. Its owner felt that the expense of repairs could not be justified and so the mill eventually had to close.

Most had already ceased operations by then. A few like Melin Cemaes and Melin Gwalchmai temporarily staved off closure by installing auxiliary engines, while Melin y Bont continued to grind corn into the 1930s using its waterwheel. Melin y Gof, Trearddur Bay had the distinction of being the last windmill to be worked by wind, not only on Anglesey, but in the whole of Wales. It operated until November 1938 when severe winter gales removed its cap and sails. Thus it finally joined all the other derelict mill towers inhabiting Anglesey's countryside, silent reminders of a bygone age.

The sails of Melin Llynon, Llanddeusant show the pronounced 'weather' typical of Anglesey's windmills.

Chapter Two

Mills, Milling and Millstones

Although nearly fifty windmills were erected on Anglesey during the eighteenth and nineteenth centuries, most closely resembled each other in design and outward appearance. Both Melin Llynon at Llanddeusant and Melin y Gof (Stanley Mill) at Trearddur Bay (Plates 1 and 2), may be regarded as being representative of the type of windmill built on the island during this period. Taken together the photographs clearly illustrate the characteristic external features of the typical Anglesey tower mill.

Four storeys in height it was constructed from rough, undressed stone and either rendered all over or, more commonly, on the windward side only. Encircling its base was a low *platform* of stone which enabled the miller to reach the sails and also prevented animals from approaching too closely. The stone tower was surmounted by a *boat-shaped cap* with gabled ends, clinker-built of horizontal weatherboarding over a wooden frame. A fringing petticoat overlapped the top of the tower and afforded protection from the weather for the wooden *rack* or *curb* on which the cap revolved.

The four *common sails* were carried on a cast-iron *windshaft* projecting through the front of the cap, tilted slightly to distribute its weight more evely and allow the rotating sails to clear the wider, lower part of the tower. Each sail comprised a lattice-like framework of laths and bars mortised to a strong wooden backbone or *whip*. This in turn was fixed to a stouter length of timber called a *stock*, almost as long as the sail itself. Behind the stock and extending about one-third of the way along the sail was another length of timber called a *clamp*. All three were joined together by iron straps and bolts and attached in a similar manner to an *iron cross* on the windshaft.

The sail bars protruded less on the leading edge of the sail with only a longitudinal *leader board* in front of the whip. This was angled to assist the flow of air over the sail cloth and help keep it flat. The cloth could be spread over the sail frame in several positions, the amount of coverage depending on the strength of the wind. A sail speed of 15 r.p.m. produced a grinding speed of 150 r.p.m. and this was considered optimal for the milling process. By grading the angles at which the bars were mortised to the whip the sail was given a twist or *weather*; this varied its pitch making it aerodynamically more efficient. On Anglesey windmills the weather was so pronounced, often as much as 30°, that it created the illusion of the sails tapering far more than they actually did.

To bring the sails into the wind the miller pulled on an *endless chain* hanging from a large wooden *chain wheel* at the rear of the cap. The chain wheel

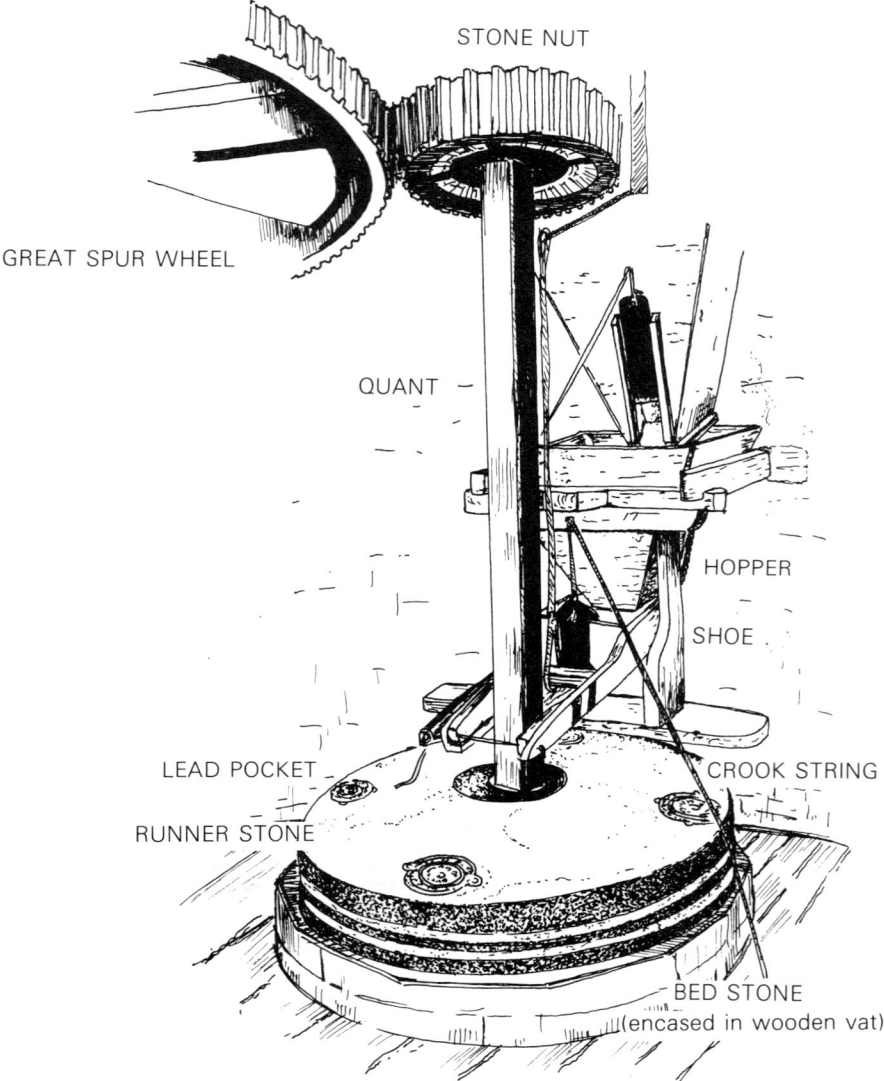

STONE NUT

GREAT SPUR WHEEL

QUANT

HOPPER

SHOE

LEAD POCKET

CROOK STRING

RUNNER STONE

BED STONE
(encased in wooden vat)

GEARWHEELS, GRAIN FEED AND MILLSTONES

connected via reduction gearing to the circular toothed rack around which the cap moved. Turning the fifteen ton cap was an arduous task for the miller, the sails having to be kept facing the wind at all times, even when the mill was not operating. When not in use the chain was secured to one of a number of wooden *cleats* which projected from the base of the tower.

Within the cap the windshaft formed an axle for the large, cogged *brake-wheel*, so called because an iron band around its rim could be contracted to slow or stop the sails. The brakewheel engaged with the *wallower*, a horizontally-set wheel mounted on an wooden upright shaft running through the centre of the mill. This was the first in a series of gears which transmitted the energy from the revolving sails to the machinery below. At the bottom of the upright shaft was the *great spur wheel* and connecting with it, on top of iron spindles or *quants*, were smaller spur wheels called *stone nuts*; these drove the *millstones*. Unlike the other gear wheels which were of iron the stone nuts had wooden teeth, usually made from applewood. Iron meshing with wood gave a smoother, quieter drive and reduced the risk of sparks; also, in times of stress the wooden teeth would shear first and it was easier and cheaper to repair a stone nut than a great spur wheel. Subsidiary machinery such as the sack hoist, flour dresser and groat machine could also be run from the main drive when required.

Sacks of grain brought to the mill from local farms were first attached to the chain of the *sack hoist* and raised through trap-doors to the second floor. Gravity was then employed to convey the grain through the various processes down to ground level again. From the second floor the grain was poured down a wooden chute to a large open container or *hopper* on the floor below. This floor was the heart of the mill and usually accommodated three pairs of millstones — one pair of *French burrstones* and two pairs of *Welsh stones*. If wheat was being ground the harder French stones were used. These were made from chert, a quartz-like rock quarried in the Paris Basin. Each millstone comprised several shaped blocks cemented together, backed with plaster and encircled by iron hoops. Grain from the hopper passed through a tapering outlet to the *shoe*, an inclined trough which directed it into the millstones. Held by a spring against the square iron quant the shoe vibrated every time it was struck by the corners of the rotating quant, so causing a regular trickle of grain to fall into the eye of the *runner stone*. This was the upper of the two grinding stones and revolved over a stationary *bedstone*. It was supported on a spindle running through the centre of the bedstone and fixed to a strong wooden beam known as the *bridge-tree*. The flow of grain to the millstones could be increased or decreased by varying the angle of the shoe. This was done by loosening or tightening the *crook-string*, a cord running from the shoe to a *twist peg* fixed in the floor.

Windmills of Anglesey

Plate 1
Melin Llynon, Llanddeusant

Plate 2
Melin y Gof, Trearddur Bay

Cut into the working faces of both millstones was a radiating pattern of *furrows* or grooves which channelled the grain to the outer edge of the stones. Each furrow had a vertical and a sloping side and as the runner stone revolved the grain was cut with a scissor-like action, the grinding process being completed by the coarse texture of the flat surfaces or *lands* between the furrows. Eventually the grinding surfaces would wear down and periodically the runner stone, weighing almost a ton, had to be lifted off the bed stone — no easy job in the cramped confines of a windmill — and the furrows and lands recut or *dressed*. Itinerant stone-dressers originally carried out this skilled task, but such craftsmen disappeared as the windmill declined in importance and the remaining millers were left to dress their own stones. When replacing the runner stone care had to be taken to ensure that it rotated in a perfectly horizontal plane otherwise it would not grind efficiently. Any fine balancing necessary was achieved by adding lead weights to pockets in its upper surface.

STONE DRESSING

The working face of a millstone is divided into sections called harps. Each harp is cut by four furrows of varying length, the longest being tangential to the eye of the stone. In the sketch the stone dresser is shaping a furrow using a mill-bill, a steel pick set in a wooden holder or thrift. To do this he sits on the edge of the millstone and rests his arm on a small sack of bran called a bist. Once the furrows are cut he will complete the dressing by chiselling a series of fine grooves into the flat surfaces between the furrows.

It was important to maintain a steady supply of grain to the millstones for if they ran dry the grinding surfaces would be worn away and contact between the stones might produce dangerous sparks. To guard against this occurrence an alarm bell hung inside the shoe. Normally this would be silent, submerged by the flowing grain, but if exposed it rang to alert the miller.

The clearance between the millstones could be adjusted to produce different types of meal. Regulating the desired gap between the stones, or *tentering*, was achieved either automatically by a centrifugal *governor* reacting to the speed of the sails or manually by raising or lowering the end of the bridge-tree on which the stone spindles carrying the runner stone were mounted.

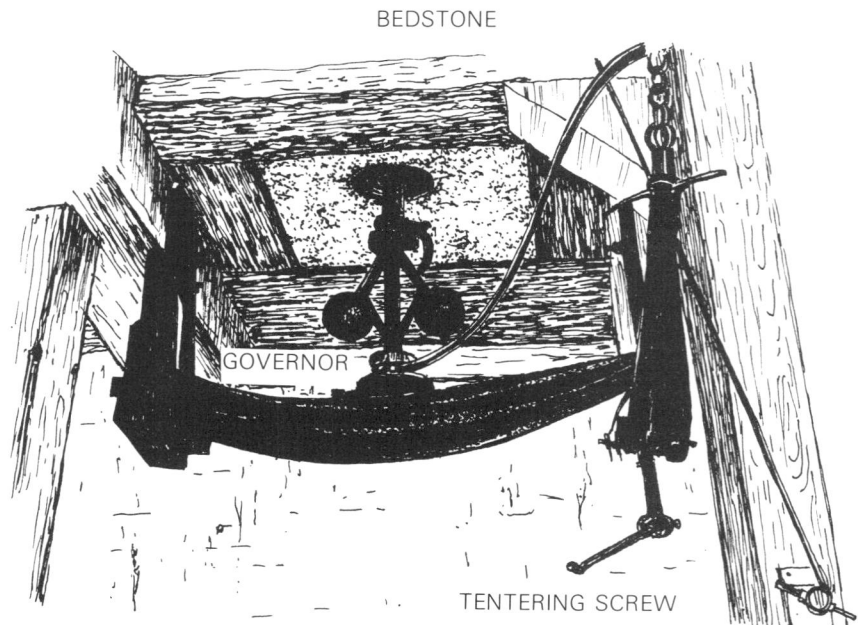

REGULATING THE MILLSTONES

The required clearance between the millstones is set before use by turning the manual screw on the tentering gear. If the wind speed increases the sails revolve faster causing the weighted balls of the governor to move outwards and pull down on a sliding collar which lowers the runner stone and prevents it from bouncing up and down. Thus grinding can continue in varying winds without the need for constant adjustments by the miller.

Mills, Milling and Millstones

Each pair of millstones was encased in a polygonal wooden box or **vat** which trapped the meal expelled from the outer edge of the stones. From here it descended through a chute to the **flour dresser** on the ground floor. The flour dresser, sometimes called a wire machine, consisted of an inclined cylinder inside which revolving brushes swept the meal against wire gauze. Only the finest flour fell through the mesh and into the wooden casing of the machine whence it was removed and bagged. The coarser middlings and bran tailed out of the end of the cylinder and into a sack.

One pair of the coarser Welsh stones was used for general grinding, the other was reserved for oats. Whereas wheat and barley were normally left to dry naturally before being ground oats were first roasted in a **kiln**, usually adjoining the mill. The oats were spread over a floor of perforated clay tiles under which a fire was lit to provide the necessary heat. Kiln roasting reduced the moisture content of the oats to a very low level enabling the husks to be easily split off. After grinding, both grain and husks passed into the **groat machine**, not unlike a flour dresser in appearance but with a four- bladed fan at one end. Inside the groat machine the red dust which had formed on the oats during roasting was brushed out through the wire mesh. As the grain and husks tailed out of the cylinder they were subjected to a draught of air from the fan which blew the light husks into a chamber called the **husk cupboard**; the heavier whole oats fell out of the machine through a spout while the lighter and poorer groats were carried to another spout further from the fan.

Unlike French burrstones which had to be imported, Welsh stones were local in origin, coming from quarries in the east of the island. Millstone quarrying on Anglesey has a long history and was already a well-established industry when construction of the windmill at Newborough began in 1303. The Bailiff's accounts confirm it was fitted with local millstones: '2 millstones bought at Mathawr half a mark. And for carriage of same from Mathawr [Benllech] to Suthkroc [Abermenai] and from Suthkroc to the site of the said mill 5s. Total 11s.8d'.

Following the sea journey to Abermenai a delay occurred in the transport of the millstones overland to Newborough, the *rhingyll* (beadle) of Menai having failed to summon the prince's bondsmen to perform this task. For this lapse of duty he was fined the sum of two shillings!

By the fourteenth century Anglesey millstones had become widely known thanks to a flourishing export trade. Contemporary records reveal ships leaving Red Wharf Bay carrying millstones to the mainland, to Ireland and even as far as the Baltic States.

Anglesey millstones were made from a conglomerate of sandstone and quartz pebbles. Considering the number that must have been produced over several centuries the deposits of this rock are surprisingly limited in extent. It outcrops in only a few places to the west of Benllech and also near Penmon. In the Middle Ages the quarries at Penmon belonged to the nearby Priory and the sale of millstones provided the monks with a valuable source of income. These quarries must still have been in operation in 1748, a letter of that date written by Lewis Morris mentioning that 'at Penmon there are several quarries of millstones of the grit kind of which great quantities are shipped off there'. Traces of them are now hard to find, the inexorable advance of vegetation having all but obscured the long-abandoned workings.

MILLSTONE QUARRYING

The sketch of an Anglesey millstone quarry has been drawn from an old photograph in Llangefni archives. It shows two workmen with picks standing next to a partly shaped millstone. A pair of finished millstones complete with iron hoops can be seen bottom right. In the ditch is a four-wheeled wagon used for transporting the stones. The pole lying across the wagon would have passed through the eye of a millstone so that it could be carried vertically, as in the diagram below.

The island's three main millstone quarries were at Pen'rallt near Brynteg, Cors Goch near Llyn Cadarn and Bwlch about one mile south-west of Benllech. Of these Pen'rallt was the most extensive and, half-hidden among the gorse, it is still possible to find finished millstones which for some reason were never

sold, perhaps because they were faulty or perhaps because there was no demand at the final rundown of the industry.

Quarrying ended just before the First World War although it had been declining
steadily for the previous fifty years, a decline closely mirroring that of the windmill. Census returns indicate that the peak of the industry had been passed by
1841, possibly also due to the growing competition from French burrstones
which by that time were being widely exported.

Interestingly, Anglesey millstones were not made by breaking off suitable
pieces of rock and cutting them to shape, but were fashioned *in situ* while still
attached to the parent rock and then fitted with iron hoops forged in the local
smithy. From the quarry the millstones were hauled away on four- wheeled,
horse-drawn wagons, usually to Red Wharf Bay for shipment. Towards the close
of the industry some were sent to Red Wharf Bay station on the Anglesey Central Railway; but rail carriage must have been short-lived, for although the
Anglesey Central Railway opened in 1867 the branch line from Holland Arms
to Red Wharf Bay was not completed until 1909.

Site of the windmill at Brynsiencyn
Reproduced from the 1901 Ordnance Survey map

Robert Dawson's map of Holyhead 1831.

Chapter Three

Lost Mills and Preservation Attempts

OVER THE LAST HUNDRED YEARS the story of Anglesey's windmills has been one of decay, dereliction and demolition. Although a survey carried out in 1929 recorded thirty seven windmills on the island in various states of repair (or, more accurately, disrepair) only four had survived in a reasonably complete condition. They were Melin yr Ogof (George's Mill), Kingsland; Melin y Gof (Stanley Mill), Trearddur Bay; Melin Llynon, Llanddeusant; and Melin Cemaes. However, when the survey was published eight years later in *An Inventory of the Ancient Monuments in Anglesey* only Melin y Gof was still in operation using wind power.

The reasons behind this neglect are not hard to understand. By the beginning of the twentieth century the age of the windmill had effectively passed. Repairs became increasingly difficult and expensive and a number of mills were abandoned following storm damage when the cap and sails were dislodged. Once the interior of a mill is no longer weathertight decay starts to set in. Exposed to the elements the wooden floors eventually rot and collapse under the weight of the millstones and other machinery, all ending up in a jumbled pile at the bottom of the tower. Sometimes this process was speeded up by deliberately setting fire to the structure to facilitate recovery of the ironwork for scrap.

Viewed from the conservation minded 1990s it is easy to condemn such actions, but in the context of the time they probably seemed quite natural. Sentiment had little practical value in the Welsh countryside and as windmills had, to all intents and purposes, outlived their useful existence what was the point of preserving them? Perhaps because they were once so numerous Anglesey's windmills were taken for granted, perhaps people were unaware they were decaying so rapidly or perhaps they just did not care. Whatever the reasons the reality is that one third of the island's windmills have now gone completely and of those that still stand most are empty shells or their crumbling remains.

Many have disappeared without trace and often map evidence serves as the only indication of a mill's original location. For example, although nothing survives of the windmill at Brynsiencyn the 1901 six-inch Ordnance Survey map shows it standing disused in a field close to the village. The sites of other lost mills can be identified in a similar way.

Windmills near the coast provided conspicuous landmarks for shipping and as such were included on several navigational charts. Thomas's sea chart of 1817 shows Holyhead possessing two windmills, an East Mill and a West Mill,

Plate 3 Dwyran

Demolished
Windmills

Plate 4 Llanddona

although only the former is marked (with a windmill symbol) on Mackenzie's earlier chart of 1775. East Mill had become Llanfawr Mill on Dawson's map of 1831 but must have fallen into disuse soon afterwards as it is not shown on Captain Beechey's chart of 1840 and occurs only as 'Mill Stump' on an Admiralty chart of 1881. No trace of this mill exists today, nor of West Mill, named Ucheldre Mill on Dawson's map. Despite appearing as Ucheldre W. Mill on the 1881 Admiralty chart it is thought to have closed some years before, its remains finally being removed in 1908 when construction of the town's Bon Saveur Convent began.

At least Melin Ucheldre is remembered in three local street names — Mill Bank, Millbank Terrace and Millbank Gardens; most have no such memorial. Occasionally, the site of a demolished mill is preserved in a house name or farm, as at Dwyran where the name 'Ty-Felin' on the gate of a cottage reveals the location of a windmill which once stood near the estuary of the Afon Braint (Plate 3).

Some windmills fell victim to housing projects. Melin Llanddona (Plate 4) was demolished just before the Second World War to make way for a Council Estate; the mill at Caergeiliog (Plate 5), listed as an empty shell in the 1929 survey, was pulled down to enable a bungalow to be built on the site; Melin Machreth at Llanfachraeth suffered a similar fate.

More recent casualties have been the two small farm windmills at Treban-Meirig and Fferam, and Melin Rhydwyn near Church Bay, demolished because it was thought to constitute a danger to local children. Most of Anglesey's remaining windmills are now scheduled as historically important buildings and anyone wishing to remove a mill must first notify the Borough Council of their intent and apply for the necessary permission. This has frustrated several farmers with dilapidated mills on their property, more than one confiding that given the chance they would gladly get rid of what to them has become a redundant and often inconvenient building.

However, the conferring of listed building status does not necessarily guarantee a windmill protection. The controversial demolition of Melin Penrhiw at Rhostrehwfa (Plate 6) by its supposed guardians Anglesey Borough Council received wide coverage in the local press. Despite being listed as a building of historical interest the mill was levelled in a clandestine operation on August 5th 1987. In the Borough Council's opinion the old tower was unsafe and had to be removed. Many villagers disagreed, and on hearing of the Borough Council's intention began a preservation campaign. Their Community Council requested a meeting to discuss the mill's future, but the Borough Council remained unmoved and early in the morning of August 5th its bulldozers reduced the

Plate 5 Caergeiliog

Demolished
Windmills

Plate 6 Rhostrehwfa

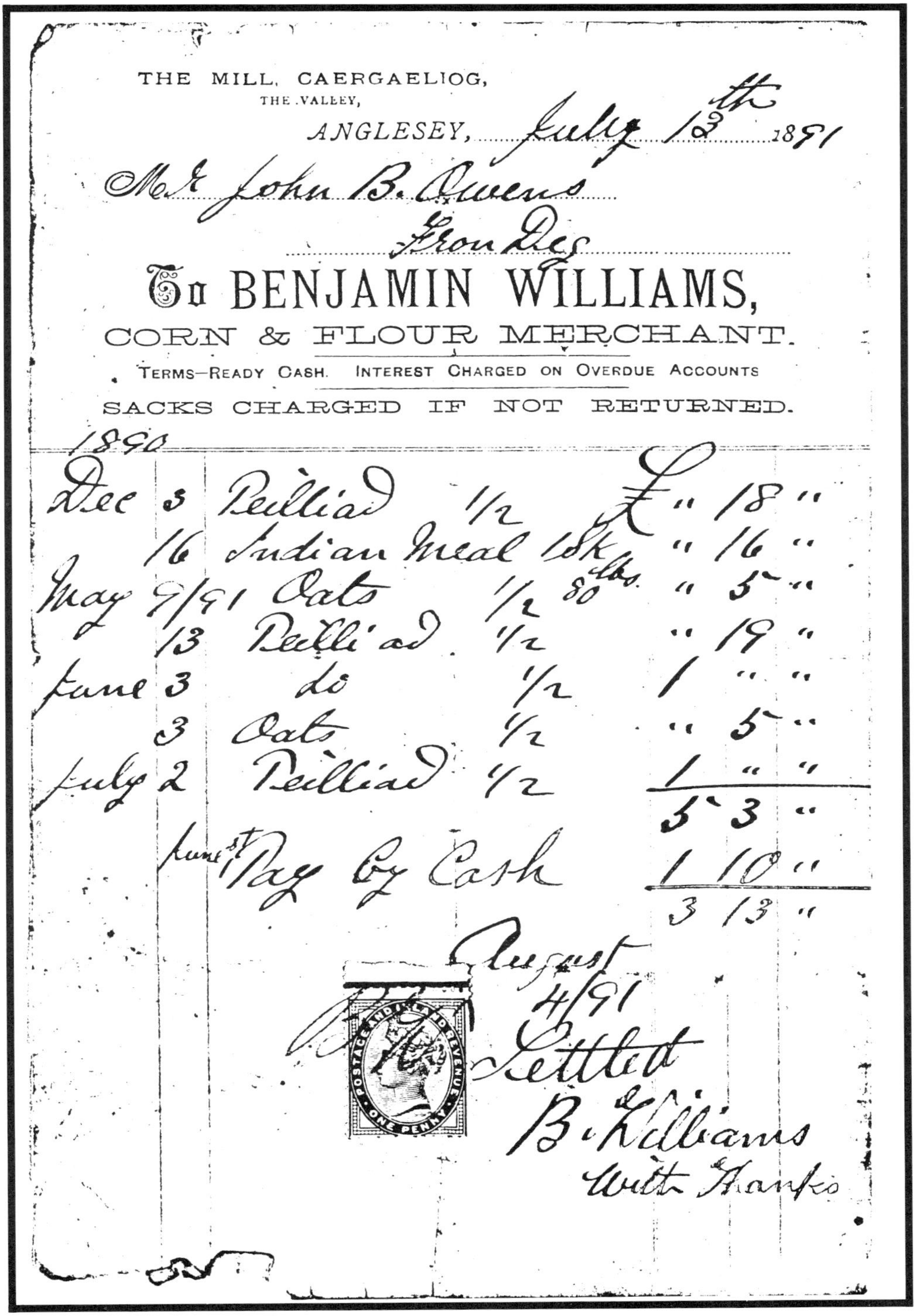

THE MILL, CAERGAELIOG,
THE VALLEY,

ANGLESEY, _July 13th_ 18_91_

Mr John B. Owens

Fron Des

To BENJAMIN WILLIAMS,

CORN & FLOUR MERCHANT.

TERMS—READY CASH.   INTEREST CHARGED ON OVERDUE ACCOUNTS

SACKS CHARGED IF NOT RETURNED.

1890

			£		
Dec 3	Peilliad	1/2		18	"
16	Indian Meal 18k			16	"
May 9/91	Oats	1/2 80 lbs		5	"
13	Peilliad	1/2		19	"
June 3	do	1/2	1	"	"
3	Oats	1/2		5	"
July 2	Peilliad	1/2	1	"	"
			5	3	"
June	Pay by Cash		1	10	"
			3	13	"

August
4/91
Settled
B. Williams
With Thanks

MELIN TYCOCH

CELLAR
Ready to fix a new Screen for Cleaning Wheat and Barley but not yet fixed.

FIRST LOFT
One Dressing Flour Machine, One Dressing Barley Meal Machine, both new Sheets and all Cylinders, two new Drums for Flour Machine and other for Barley Meal ditto, and three new leather Straps and two new wood palings and iron pallet and three new iron hangers and three new Werthyd [axles] and three new bridge [bridge trees].

SECOND LOFT
One new pair of French Grinding Stones, two new pairs of Bwlch Gwyn Stones for grinding Barley and Oats, three new iron cribun [quants] and three new iron hangers, and three new iron Spurnut [stone nuts] and one Spur Wheel and one new iron cog wheel and iron pallet.
THIRD LOFT

One wood new Drum for hoisting Sacks and one iron chain for ditto. Three wood Bins to hold the Corn before grinding. One new leather Strap for hoisting Corn.
FOURTH LOFT

One new Pallet [upright shaft] made from the best Pine and one iron Spur Wheel [brake wheel] and one Counter Wheel [wallower] and new iron turn Gear and one iron Wheelhole and Turning Gear and two new Wood Rings and all brass (?) steps from the lower part to the top of the new Mill.

OUTSIDE
Four New Wood Wings [sails] made of the best Pine and two iron cross, one outside and other inside, and one iron Pallet [windshaft].

mill to rubble, much to the indignation of local people who felt they had been denied the opportunity to put their case.

Two lost windmills which, for different reasons, merit closer examination are Melin Tycoch, Dulas and Melin Machreth, Llanfachraeth.

Melin Tycoch, the last corn windmill to be built on Anglesey, experienced an unusually brief working life. It was erected in 1862 by Hugh Jones, less than eighty yards from a long-established watermill of which he was tenant. The watermill had been in existence since 1771, a document of that date recording that 'Hugh John Hughes of Tyfoncoch . . . hath agreed to erect and build . . . at his own expense one Water Corn Grist Mill . . . upon or near the site of a mill that formerly heretofor stood' (an allusion to Llysdulas Mill which was mentioned in the 1352 Extent of Anglesey).

In December 1862 Hugh Jones signed a lease with his landlord Hugh Robert Hughes of Kinmel Park, Denbighshire. He agreed to pay Hughes an annual rent of £5 for the property, described as consisting of 'All that Water Corn Mill with the Machinery Dams Weirs Sluices and Watercourses thereunto belonging Together with the Reservoir Pools and Races and two pieces of land thereto belonging and adjoining. And also all that new Wind Corn Mill recently built thereon by the said Hugh Jones together with the Stones Mill Gear and Machinery lately placed and fixed therein'.

The terms of the lease required Jones to 'keep in good and substantial repair the said Mills . . . and also all the millstones mill gear and machinery in or about the said premises'. Included with the lease was a plan of the property and a detailed description of the newly built windmill.

Hugh Jones died in 1880 and to all intents and purposes his windmill died with him. There is no evidence that his widow Catherine took over the mill, indeed, a survey of 1887 revealed that it had by then fallen into disuse, probably having closed soon after Hugh Jones' death.

In hindsight it seemed an ill-conceived venture. Given the declining fortunes of arable farming on Anglesey during the second half of the nineteenth century building a new windmill, even in this remote part of the island, must have constituted something of a gamble. Perhaps Hugh Jones was concerned about the reliability of the water supply to his watermill and saw the windmill as insurance against this eventuality.

Today no trace remains of either the windmill or the watermill and only documentary evidence betrays the fact that they ever existed.

Of all the windmills which have disappeared probably the greatest loss has been that of Melin Machreth, Llanfachraeth. Second in size only to Amlwch Port Mill it served a wide area, with deliveries as far afield as Bryngwran, Gwalchmai and Aberffraw. An early description of Melin Machreth appears in

PLAN

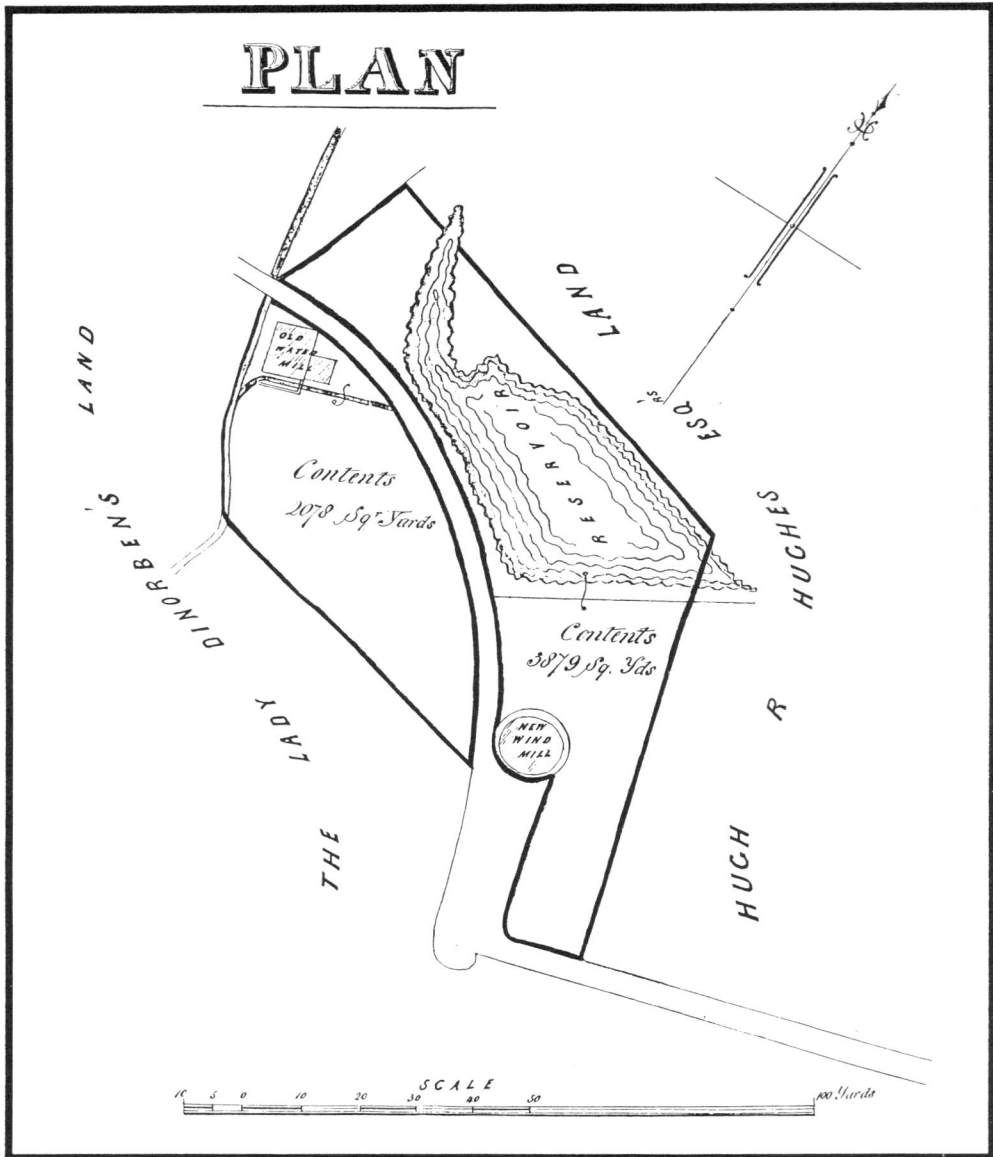

Plan of mills at Melin Tycoch, Dulas

the book *Mynydd-y-Gof* by David Roberts, a local doctor whose father owned shares in the mill. As a child (in the 1830s) he recalled being sent to Llanfachraeth with messages for the miller. These visits to the mill, usually in the company of his brothers, were eagerly anticipated.

'Our ears were strained to hear its sound as we approached. The first intimation that reached us was a deep mysterious breathing. Then after passing a little bend in the road a hurried sighing and sobbing made itself felt. A quick turning brought us in full sight of the mill itself with its white sails groaning and flapping and as if marching in pursuit or escaping from some unseen enemy. The big white tower with its huge wings seemed to my imagination like a giant.

The miller knew us well. He was a preacher as well as a miller, by name William Williams. He had a peculiar voice, husky as if some oatmeal had stuck in his throat. He was very kind and showed us the great grinding-wheel and another one again on the next flat, and would carefully conduct us to the cage attached to the cap. When the wind was high this cage shook a good deal which frightened us not a little.

But the view was grand . . . Holyhead mountain and town seemed close to us. Besides the church and churchyard below, the creeks and channels that the sea made in the sandbanks bordering the little estuary of the Alaw lay beneath and around us.'

For much of the last quarter of the nineteenth century Melin Machreth was worked by Rice Rowlands, one of seven brothers involved in milling. Robert ran Melin Llynon at Llanddeusant until its closure, John and Richard were at Rhydwyn, William at Cemaes, Rowland at Gaerwen, and Owen at Dulas and Llangefni. On Rice Rowlands' retirement his son, Robert Buckley Rowlands took over Melin Machreth and continued to operate it until 1917 when the cap was dislodged in a violent storm. Prior to this event the mill had been running with only two sails. Business had not been good and the owner's reluctance to replace the missing sails suggests that he harboured doubts about the future viability of the mill. By then it had been reduced to crushing oats and Indian corn for animal feed, and although retaining its capacity to grind wheat and barley there had been little demand for this service for several years.

In 1918 Melin Machreth (and adjoining land) was sold as one of the lots in an auction of part of the Garreglwyd and Berw Estates. Although described in the auction catalogue as 'The Powerful Four-Winged Mill' Melin Machreth's grinding days were over. Soon after closure one pair of its millstones was removed to Valley and had a new lease of life in the steam mill owned by the Gardner family, close friends of the Rowlands.

The causes of Melin Machreth's demise were presaged in Dr. Roberts' book,

Parish of Llanfachraeth.

Lot 24

ABER ALAW MILL, LAND, COTTAGES, &c.

In the Occupation of Mr. R. B. ROWLANDS, now let at an Annual Rental of £37.

No. on Plan.	DESCRIPTION.	Quantity.		
		A.	R.	P.
87	Wind Mill, House, Outbuildings, Etc.	0	3	22
86	Cae Capel	0	3	3
88	Tanfelin Cottage and Yard	0	0	12
83—84	Tai Ysgoldy Cottages and Gardens	0	0	8
	Total Area	1	3	5

AN IMPORTANT

Mill Property Cottages and Land,

Most centrally situated on the main road leading through the
Village of Llanfachraeth.

The Powerful Four-Winged Mill,

Strongly built, containing Corn and Flour-Dressing Machinery with ample storage
accommodation in the Mill proper and adjoining commodious Granary.

TWO QUILLETS OF LAND AND FOUR COTTAGES,

Two Large Stables with Lofts, Store House, Office, &c., form a very valuable adjunct
to the Mill, which has a widely distributed connection.

published in 1905. He wrote 'The old windmill still stands but the bulk of the flour trade is done by steam or imported from other places'.

Rice Rowlands must have been aware of the changes occurring for he astutely built warehouses at the Valley end of the Stanley Embankment to store imported flour shipped from Liverpool and unloaded on the pier at Valley Cob. A Fordson Steam Wagon was purchased in 1914 to draw the mill carts and facilitate the speedy distribution of the flour. In 1921, a year after his father's death, Robert Rowlands sold the flour business to the North Shore Mill Company, but still continued to farm at Aber Alaw.

Melin Machreth was eventually demolished and its stone incorporated into the house which now stands on the site. However, continuity with the Rowlands family has not been broken for the house belongs to the Reverend John Rowlands, son of Robert and grandson of Rice.

Not everyone remained indifferent to the plight of the island's windmills. The closure of the mill at Trearddur Bay prompted what was probably the earliest, if ill-fated, attempt at preservation on Anglesey. Known locally as Melin y Gof the windmill was a prominent and frequently photographed feature of this quiet but popular holiday resort. Much public concern was aroused over its closure and its owner, the Honourable Lyulph Stanley, offered to donate the mill, its site and right of way to local trustees on condition that steps were taken to have the building repaired and maintained. It was estimated that £600 would be required and an appeal was launched in the summer of 1939. Unfortunately, the outbreak of the Second World War caused the project to be abandoned and the money that had been collected was donated to Red Cross funds.

After the war building materials were in short supply and priority had to be given to the reconstruction of houses, factories, hospitals and schools. Several years were to elapse before there were signs of a revival of interest in windmill preservation at an official level. In June 1953 Anglesey's Planning Committee, concerned about the possible loss of one aspect of the island's heritage, resolved that 'The Welsh Church Acts Committee be asked to consider the preservation of the fabric of a typical windmill in the County'.

To assess the condition of those windmills still extant the County Council carried out a survey which, in retrospect, turned out to be less than thorough and contained a number of mistakes. For example, the entry for Melin Penrhiw, Cerrigceinwen stated 'No evidence of the existence of this mill was found', whereas in reality it could clearly be seen from the B4422 at Rhostrehwfa until its demolition in 1987. This error was of little consequence, but another later proved to be more significant. The entry read 'George's mill, Trearddur Bay. This mill is in a fairly good state of repair. The door is in good condition and all the windows have been boarded up. There are no sails or mill machinery in existence'. The description referred to Melin yr Ogof, or George's Mill, Kingsland which was, and still is, the only windmill on Anglesey to retain all its original machinery except the cap and sails. The official who did the survey obviously failed to go inside!

Ignorant of the true condition of Melin yr Ogof the County Council decided to approach the owners of Melin Cemaes and Melin Llynon with a view to purchasing one or both properties. Comprehensive surveys of the two mills were commissioned from Rex Wailes, the noted windmill authority.

His report on Melin Cemaes concluded that apart from the tower virtually the whole mill would have to be rebuilt and new machinery installed, and estimated the cost of this work to be £3,000–£3,500. Although he found the condition of Melin Llynon to be 'extremely bad' it was still somewhat better than that of Melin Cemaes and consequently the cost of restoration would be slightly less, around £2,500–£3,000, 'having regard to the facts that the work is beyond the capacity of a builder and the nearest millwrights capable of undertaking the work are in Lincolnshire'.

Not surprisingly, the County Council opted for the cheaper alternative and agreed to ask the owner of Melin Llynon if he would be willing to sell the mill at a nominal figure if alternative storage accommodation was provided. This he was reluctant to do.

Meanwhile, at the request of Rex Wailes, the mill had been visited by representatives of R. Thompson and Son, a firm of millwrights from Alford in Lincolnshire, who quoted a price of £3,940 for its renovation. Faced with this estimate, plus another £375 from a local builder for additional work, and a further £100 for accommodation in lieu of storage, the County Council began to have second thoughts. Nevertheless, its report was submitted to a meeting of the Welsh Church Acts Committee in October 1955. Unhappy about a potential bill of £4,415 the Committee resolved that 'the proposed restoration of the windmill be not proceeded with and that no further action be taken in the matter'.

Perhaps if Anglesey County Council had been aware of the excellent condition of Melin yr Ogof at Kingsland it would have been more successful, the owner having no objection to his mill being restored. As well as being cheaper it would also have satisfied a suggestion by Thompsons that, as Melin Llynon would require virtual rebuilding, it might be worth considering the restoration of a windmill more accessible to the general public.

Interest in Melin Llynon was briefly rekindled in 1966 following the death of its owner. Thompsons estimated that costs would have doubled in the eleven years since their initial survey, and the County Council applied to the Secretary of State for Wales for financial assistance. He was willing to make a grant of

75% up to a maximum of £7,013 (based on the Ministry architect's report which assessed the cost of restoration at £9,350) providing that members of the public could have access to the interior of the mill. However, for a number of reasons the project was again shelved.

October 1970 saw the County Architect once more writing to Thompsons for yet another estimate. They replied 'We are pleased to hear that there is some thought of restoring the old mill' and considered that a 50% increase on their 1966 quotation (ie. a figure of £11,820) was not unreasonable. But despite Melin Llynon being in a very derelict state the new owner was still not interested in selling. Neither was the owner of Melin Cemaes. During these protracted wranglings another possible candidate for restoration, Melin-y-Gof at Trearddur Bay, had been converted into a house, planning permission having been given in spite of the pre-war attempt to preserve it.

In 1971 Anglesey County Council at last turned its attention to Melin yr Ogof, Kingsland. On its behalf Rex Wailes surveyed the mill and, rather ironically, suggested the acquisition of the turning gear and curb from Melin Llynon. He considered that it could be adapted to fit Melin yr Ogof but without it the cost of restoration would be prohibitive. In the event the idea was dropped, either because the owner of Melin Llynon would not part with his ironwork or because the County Council was put off by the high cost of new castings.

During the next four years Melin yr Ogof changed hands and in 1975 its new owner applied for planning permission to convert the mill into a five- bedroomed house. This would have meant the destruction of its machinery. Objections were raised and following a press campaign mounted by co-author George Lees support was received from Cledwyn Hughes (now Lord Cledwyn of Penrhos), Rex Wailes and the S.P.A.B., the Historic Building Council for Wales, the Council for the Preservation of Rural Wales, the R.C.A.H.M., St. Fagan's Museum, the Director of European Heritage Year 1975, and many Anglesey residents. All to no avail, the Local Planning Authority granting planning permission and listed building consent for the mill's conversion.

Three years later, on 2nd November 1978, the Holyhead Mail carried the following advertisement . . .

BY INSTRUCTIONS OF MR. T. H. ROWLANDS:
TO BE SOLD BY PUBLIC AUCTION
on TUESDAY, 28th NOVEMBER, at 3.30 p.m.
in THE VALLEY HOTEL
THE DESIRABLE FARM with approx. 57 ACRES
known as
FELIN LLYNON, LLANDDEUSANT
including the Ancient Windmill and 4½ acres
of land.

This aroused great interest among the residents of the nearby village of Lland-deusant and by 23rd November, 860 people had signed a petition requesting Anglesey Borough Council (which had replaced the County Council following Local Government reorganisation in 1974) to purchase and restore Melin Llynon. On 28th November it paid £10,000 for the windmill, the old granary and 4½ acres of land. The story of Melin Llynon's eventual restoration is told in Chapter Four.

Meanwhile, although Melin yr Ogof had received planning permission for conversion to a dwelling, the owner had made little attempt to convert the mill, nor was he even carrying out essential repairs to keep it weatherproof. In 1984 the newly formed Welsh Mills Society wrote to Anglesey Borough Council expressing its concern about the future of the mill and the deterioration of the structure as a result of lack of maintenance.

In 1986 Melin yr Ogof's owner put the mill up for sale at £24,000, the property being described as 'a superb example of an Anglesey Mill complete with internal workings having planning permission for conversion to dwelling with five bedrooms'. The Borough Council was urged by the Welsh Mills Society to consider acquiring the mill for restoration, perhaps as a static exhibit to compliment Melin Llynon, but despite an active press campaign and a petition containing over 1300 signatures it decided against purchase.

Throughout 1987 the efforts of the Welsh Mills Society to save Melin yr Ogof received much media attention and early the following year the mill was bought by Stuart Ward, a member of the W.M.S., with the intention of restoring it to full working order. However, the Planning Committee of Anglesey Borough Council refused permission for total restoration on the grounds that the height and fixed nature of the proposed sails would create a potential hazard and noise disturbance to the residents of the adjoining properties. An appeal was lodged against the decision claiming that the sails were essential to drive the machinery as the owner wished to work the mill occasionally and demonstrate it to interested parties. Also it was necessary to turn the sails from time to time to prevent the machinery from seizing up. As yet the appeal has not been heard and the future of Melin yr Ogof remains uncertain.

What of the future? With the exception of Melin yr Ogof there are, sadly, no more windmills on Anglesey capable of being restored. Even if there were, current planning legislation seems antipathetic to such projects. Anglesey is not unique in this respect, Wales in general having a poor record as far as windmill restoration is concerned. Although around 150 windmills have been recorded throughout the Principality only one, Melin Llynon, has been restored to working condition. On Anglesey there is still some scope for watermill restorations,

Lost Mills and Preservation Attempts

but recent attempts to renovate Melin Bodowyr near Brynsiencyn and Melin Dulas have run into difficulties, the former from an uncooperative landowner and the latter from restrictive conditions laid down by the island's Planning Authority.

Plate 7 Melin Llynon in 1971 showing clearly the result of over forty years of neglect.

Chaper Four

Rescue and Restoration

'Past the little white cottages and the grey farmhouses we went; past the little grey churches with their tiny bell towers at the west gable, and the old sail-less towers of windmills (plain evidence of the amount of corn which must have grown in Anglesey's fertile soil years ago), until we came to Llanddeusant where there is a windmill, the only one on the island I believe, which still retains its sails. There it stood, on high ground to the west of the village, its four great arms fixed and immovable, and the slats of the sails hanging loose and derelict, the playing perches of Jackdaws and the look-out posts for Kestrel and Carrion-crow. Daylight showed through the curved timbers of the roof, through which the weather penetrates to the interior timbers, hastening their decay and destruction.

Though the mill had not worked for many years the dust of the flour still seemed to linger about the interior and it did not require much effort to imagine the great cog-wheels and stones turning once more. But decay has gone too far I fear and the grand old mill slowly follows the fine old craftsmen who made it, into oblivion. We left it wondering sadly if, on our next visit, it would still be inviting the winds with arms akimbo, or whether we should find a limbless stone tower similar to all the other mill-towers in Anglesey.'

from *Shorelands Summer Diary* by C.F. Tunnicliffe

CHARLES TUNNICLIFFE, the distinguished wildlife artist, moved to Anglesey in 1947 settling at Malltraeth on the estuary of the Afon Cefni. *Shorelands Summer Diary*, an illustrated account of his first six months on the island, was published five years later. Tunnicliffe's gloomy forecast for Melin Llynon's probable fate seemed destined to be confirmed over the ensuing years as various schemes for its restoration all foundered, usually through lack of money or the owner's unwillingness to sell. During this time the mill's condition continued to deteriorate.

Melin Llynon had been chosen by Rex Wailes in 1929 for inclusion in *An Inventory of the Ancient Monuments in Anglesey* as 'one of the very few windmills which are still in working order'. He wrote a detailed description of the mill and a comprehensive account of the workings of its machinery. However, by the late 1930s Melin Llynon had fallen into disrepair and in 1954 it sustained serious damage when a storm removed the cap and much of the sails, leaving only the cast-iron windshaft and cross in place.

Plate 8
A hired crane begins the
task of removing material
from the derelict tower.

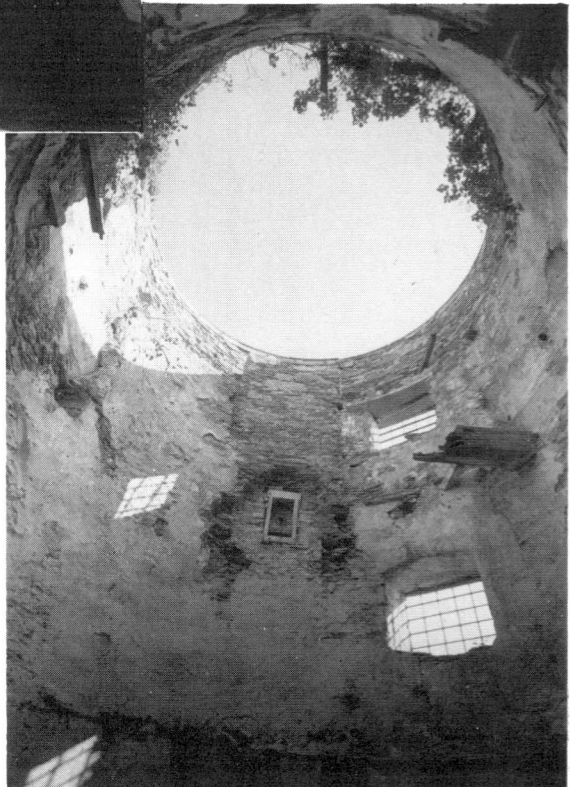

Plate 9
Sunlight streams into the
roofless, now empty shell
of the mill.

Rescue and Restoration

In 1973 Tony Parkinson visited the mill on behalf of the Royal Commission on Ancient Monuments and his report did nothing to dispel the fears for its future expressed by Charles Tunnicliffe twenty one years earlier. Parkinson found that 'While the condition of the stonework appears to be still good, the loss of the cap has meant that damp and rot has spread throughout the internal wood-work. The breast-beams appear to have given way, allowing the sails to settle and their tips to rest on the ground . . . the entire weight of the wallower and spur wheel, both of cast iron, is held by a single massive beam strapped to two cross-beams bedded in the masonry; the support beam is cracking. Almost all the floorboards have gone, and one of the two first-floor beams has broken, throwing the pair of wheat-stones to the floor upside down. The groats stones are slipping and may also collapse'.

Surprisingly, much of the rest of the machinery seemed to be in reasonable condition, but Parkinson considered that it would not be possible to save the mill 'unless action is taken within the next year or two'.

In the event it was to be five years before anything could be done. In 1978 the owner of Melin Llynon, Thomas Rowlands, retired from farming and decided to sell the mill. Responding to petitions from several organisations Anglesey Borough Council, to its great credit, agreed to purchase the mill and restore if 'for future generations of islanders to enjoy and learn about their heritage'.

But fifty years of neglect had taken their toll and Melin Llynon was now in a very dilapidated state. The one sail that remained hung precariously from the rusted windshaft, while inside the floors had collapsed and machinery, millstones and rotten timbers lay in a jumbled heap at the bottom of the tower. The mill was in such a dangerous condition that a sign warning visitors to keep out had to be erected.

Responsibility for transforming this scene of dereliction into a fully operational windmill fell to Frank Stead, Assistant Planning Officer for Anglesey Borough Council, overall supervision of the project being in the hands of Reg Powell, Director of Planning. Confessing to having very little knowledge or experience of restoring windmills Frank Stead was told 'imagine it was your own'! He after-wards admitted that the restoration had been rather like trying to solve a jigsaw puzzle with some of the pieces missing and gave most of the credit for its successful completion to the expertise of the millwrights R. Thompson and Son of Alford, Lincolnshire.

Following its purchase by Anglesey Borough Council Melin Llynon was inspected by Jim Davies and his son Tom (of Thompsons) and Rex Wailes who had written the original description of the mill fifty years before. Tom Davies vividly remembers that visit: "When we first set eyes on the mill and looked though the ground floor doorway our hearts sank at the sight before us". His

Plates 10 and 11
Salvaged machinery is set aside to be cleaned and repaired. The woodwork has rotted but the cast-iron windshaft and cross (above) and the millstones and governors can be refitted later.

father, however, relished the challenge offered by Melin Llynon saying ''We haven't been beat yet and we're not starting now''! When asked by the Council if it would be feasible to restore the mill Jim Davies replied, hand on heart, ''If we have got the time and you have got the money then it can be done''.

Having decided to go ahead detailed plans were then drawn up based on an earlier survey by Thompsons in 1954, Rex Wailes' notes and photographs, and measured drawings of the mill preserved in the National Monuments Record for Wales at Aberystwyth. These were to prove invaluable in ensuring the accuracy of the replacement parts.

The first task was to clear out the contents of the mill and trying to salvage any items that could be reused. In what Tom Davies called the *fishing exercise* a 45-ton crane carefully lifted out the rotten timbers and pieces of machinery, a job which often saw him sitting on the ivy-covered rim of the tower directing operations below. Once emptied, the tower was fitted with a temporary corrugated-iron roof to prevent further deterioration. None of the wood was reusable but the millstones and most of the metalwork had survived in a reasonable condition. The windshaft and cross, brakewheel gear segments, turning gear rack and gears, wallower, great spur wheel, stone nuts and parts of the sack hoist and governors were sandblasted clean, repaired where necessary and painted ready to be refitted at a later date.

Clearing out the tower and removing the ivy embedded in the stonework revealed a number of structural cracks which had to be made good before work could continue. Because the joists supporting the floors had rotted it was necessary to cut several holes in the side of the tower through which new timbers could be inserted. Finding the right type of wood proved a problem but eventually, after an extensive search, suitable beams of greenheart were located in one of Liverpool's old warehouses.

Rotten lintels and broken slate sills had to be replaced and new iron window frames cast utilising the best of the original frames as a pattern. The preparatory work was completed by giving the external wall a protective coat of cement render.

For financial expediency Anglesey Borough Council decided to phase the millwrights' work over a three year period, from 1981 to 1983. Thompsons happily agreed to this as it meant they would not have to neglect their other customers. In March 1981 they submitted their estimate for Stage One of the restoration work; this included making and fitting the wooden curb on which the cap and sails would eventually revolve, the frame for centring the cap, and the upright shaft to carry the wallower and great spur wheel. Most of this work could be carried out at Thompson's premises in Alford, although it was first necessary to visit the mill to make appropriate templates and bring back all the

Plate 12
Scaffolding envelops the tower as structural repairs get under way.

Plate 13
The freshly rendered tower, complete with replacement windows, emerges from its cocoon of scaffolding.

Plate 14
To facilitate the smooth turning of the cap twelve mild steel rollers were added to the curb, a departure from the original design in which the cap revolved only on well-greased wooden blocks.

Plate 15
Suspended in mid-air the oak cap frame is slowly lifted to the top of the tower.

Plate 16
Thompson's 'piece de resistance' — Melin Llynon's new oak and elm brakewheel; behind is the cleaned and painted iron windshaft and cross.

Plate 17
The assembled windshaft and brakewheel are carefully manoeuvred into position.

Plate 18
Standing on the cap frame Tim Farnsworth and Denis Moore supervise the installation of the windshaft and brakewheel.

Plate 19
With the turning gear in place Jim Davies begins work on the framework of the boat-shaped cap.

Plate 20
Working thirty feet above the Anglesey countryside Tom Davies and Denis Moore need to have a good head for heights.

Plate 21
The cap nears completion as all four millwrights work together to fix the last of the overlapping planks.

Plate 22
Inside the finished cap the windshaft, brakewheel and turning gear are protected from the elements.

Plate 23
Jim Davies gives the guard rail at
the rear of the cap a protective
coat of creosote.

Plate 24
In Alford Tim Farnsworth and Denis
Moore assemble a pair of sails for
Melin Llynon.

metal parts needed to complete the work. Suitable oak was obtained from Somerscales of Keelby near Grimsby, a wood yard favoured by Thompsons because of its high quality stock and wide selection of curved-grain timber ideal for their requirements.

Back in the workshop the curved sections of the curb were planed into shape before being bolted together, the mortised and tenoned centring frame was made up around a cast-iron ring bolted to the frame, and the upright shaft fashioned into the requisite octagonal cross-section. Once everything fitted together satisfactorily it was dismantled, transported the 265 miles to Anglesey, and reassembled at the mill.

Initially, one of the main problems facing the millwrights was finding their way to and from Llanddeusant through the maze of unfamiliar narrow lanes. Tom Davies confessed that ''on one occasion we left the mill to find our way back to Lincolnshire and after covering ten miles came across a road sign saying 'Llanddeusant 1½ miles'!'' During the restoration work they usually spent the summer months on site, staying for two weeks at a time to minimise the travelling.

Stage Two, in 1982, began with the raising into position of the heavy oak cap frame. Fixed to its underside was a curb ring designed to run on the blocks and rollers already in place on the tower rim. Before the reconditioned windshaft and cross could be refitted new brass neck and tail bearings had to be cast to replace those worn down by years of service. Strong bearings were essential for they supported the weight of the windshaft and helped to facilitate its smooth rotation. Of the original brakewheel only the iron cog segments remained, all wooden parts having rotted beyond repair. Basing their measurements on the surviving ironwork Thompsons constructed a new brakewheel, using elm for the curved rim sections and oak for the bracing cross-arms. Windshaft and brakewheel were then lifted into position and the latter adjusted to ensure a correct mesh with the wallower at the upper end of the upright shaft.

A start could now be made on constructing the mill's boat-shaped cap which, when finished, would enable work to continue within the tower under cover. Specially selected curved oak spars formed a ribbed framework over which planks of deal were bent and clamped before being nailed into place. The installation of a new chain wheel and the fitting of a guard rail to three sides of the turning gear staging at the rear of the cap completed the millwrights' work for the year.

As winter approached and the millwrights returned to Lincolnshire the carpenters moved in. Sheltered by the weatherproof cap William Ellis Jones and Hugh Jones of Llangefni were able to lay the remaining floors (including

Plate 25
The indispensable crane raises the
last of Melin Llynon's four sails
into place.

Plate 26
Restoration complete.

KEY

1 Sail
2 Iron Cross
3 Windshaft
4 Brake Wheel
5 Great Spur Wheel
6 Upright Shaft
7 Wallower
8 Stone Nut
9 Quant
10 Grain Hopper
11 Shoe
12 Runner Stone
13 Bedstone
14 Stone Casing
15 Tentering Gear
16 Governors
17 Grain Chute
18 Cleat
19 Chain Wheel
20 Turning Gear
21 Cap Centring Frame
22 Sack Hoist

SECTION THROUGH MELIN LLYNON
Prepared from a drawing kindly supplied by Anglesey Borough Council Planning Department.

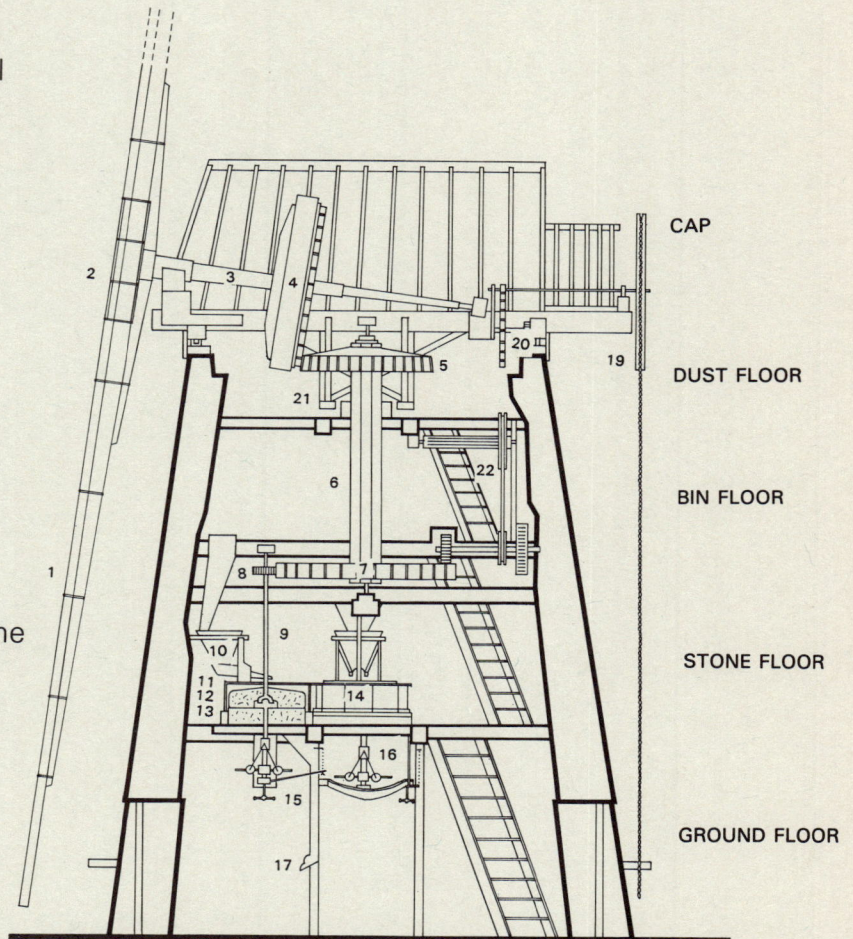

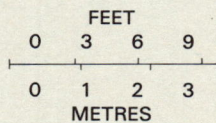

CAP

DUST FLOOR

BIN FLOOR

STONE FLOOR

GROUND FLOOR

FEET
0   3   6   9

0   1   2   3
METRES

trapdoors for the sack hoist), fit the connecting steps and make and hang the external doors.

Spring 1983 saw the start of Stage Three, the final stage of the project. This involved the installation of the mill's internal machinery and the fitting of four new sails. First the wooden support frames for the millstones were fixed in place together with the refurbished ironwork including spindles, tentering gear and governors. Next the three stone nuts, regeared with fresh cogs of beechwood, were trued up to the great spur wheel at the lower end of the upright shaft. Two of the three pairs of millstones were then dressed and set up to full working order, the third pair being left open to public view. Working closely with the millwrights the carpenters expertly made and fitted the millstones' wooden casings, the grain spouts, hoppers, shoes and flour chutes.

Meanwhile, in Thompson's cramped yard at Alford, four new pitch pine sails, each over 36 feet long and weighing 13 cwt, were taking shape. By early August they were finished and on their way by road to Anglesey. On August 8th, as the millwrights began the task of fitting the sails, the entire village of Llanddeusant seemed to be making its way to the mill to witness the final act of the drama. Sean Hagerty, who took most of the photographs used in this chapter, recalled the scene: 'By the time the crane was lifting the first sail into place, half the village was present and the other half followed shortly. What began as a routine operation turned into a gala performance. Around midday vacuum flasks and sandwiches were produced and a carnival atmosphere prevailed, each stage of the operation being applauded by the appreciative audience'.

With the sails fitted, external work on the mill was effectively complete and only the finishing touches remained. The interior walls were plastered and the ground floor relaid with old millstones set in the concrete; both internal and external walls were painted white and the cap given a final coat of creosote.

Melin Llynon's restoration had cost £120,000, considerably more than the £2,500—£3,000 estimated by Rex Wailes in 1954! Finance for the project came from Anglesey Borough Council with money from the Shell Oil Revenue Fund, and a generous grant from the Historic Buildings Council for Wales.

Thompsons regarded the project as one of the best they had ever worked on, and greatly appreciated the hospitality shown to them by the local people, particularly the Hughes family who provided the four millwrights with a 'home from home' during their stays on the island.

On May 11th 1984 Melin Llynon, resplendent in its white coat, was officially opened by the Major of Anglesey, Councillor T.D. Roberts. The Marquess of Anglesey gave an address and a steady breeze once again set the sails in mo-

Plate 27
Official guests at the opening of Melin Llynon, May 11th 1984. From left: F.G. Stead (Assistant Planning Officer), J. Meirion Davies (Chairman Planning Committee), E.L. Gibson (Chief Executive), T.D. Roberts (Mayor of Anglesey), Marquess of Anglesey (Chairman of Historic Building Council for Wales), Ann Lloyd Jones (Deputy Mayor), W. Evans (Deputy Director of Planning), R.F. Powell (Director of Planning).

tion after an interval of sixty years. Jim Davies, Tom Davies, Tim Farnsworth and Denis Moore stood quietly in the background while the dignitaries made their speeches, content in the knowledge that the success of the project had been largely due to their skill, hard work and craftsmanship.

Later that year a full-time miller, Phil Williams, was appointed and Melin Llynon's sails now turn most days when the weather permits. Safety considerations for visitors precluded the refitting of a flour dresser inside the mill, but one has recently been installed in the old kiln house next to the car park enabling wholemeal flour to be produced for sale to the public. The former granary adjacent to the mill has been converted into a tearoom and shop in the capable hands of Dilys Hughes, the building also housing an interpretative display and toilets.

Since its opening Melin Llynon has become one of Anglesey's most popular tourist attractions, receiving thousands of visitors annually including many from abroad. It seems set for a successful new lease of life.

Gazetteer

THE VISIBLE REMAINS OF THIRTY ONE WINDMILLS can still be seen on Anglesey, ranging in condition from truncated or partially collapsed towers to a fully restored and working windmill. Most are empty shells, although one retains its internal machinery and six others have been converted into dwellings.

Historical information concerning individual mills is equally varied, a fact reflected in the coverage given in the following pages. For a number of reasons the majority of Anglesey's windmills closed long before their counterparts in England; most did not survive the First World War and several had already been abandoned by the end of the nineteenth century. The few that continued to work after 1918 are generally much better documented, one or two even remembered in operation by elderly inhabitants.

To avoid confusion — several mills being known by more than one name — the entries are arranged in alphabetical order of Community, an administrative unit employed for Local Government purposes. Often a community boundary closely corresponds with that of a parish but in some cases, particularly in sparsely populated areas, a community may comprise two or more parishes.

Of the thirty one windmills still extant all but one were corn mills, processing the grain brought to them from local farms. The exception stands on the summit of Parys Mountain in the north-east of the island and was a windpump built to help remove water from the underlying copper workings. Although located near Amlwch it is not included with the other windmills in that community but, because of its different function, is considered separately at the end of the gazetteer.

Originally, it had been intended to illustrate each entry with a photograph showing the windmill in working condition but, regrettably, this has not been possible, early pictures of certain mills having proved unobtainable. Recent photographs have been included in their place, and provide an interesting comparison as they more accurately reflect the present state of most of Anglesey's remaining windmills.

The Ordnance Survey map reference for each windmill is given to aid location and those readers who wish to search out particular mills will probably find Sheet 114 (Anglesey) of the 1:50,000 Landranger Series most useful. All but a few of the mills described overleaf can be viewed from public roads or paths, but as they are usually on private land permission should first be sought before closer examination is attempted. Some are in a dangerous condition and should only be entered with care.

A map at the end of the gazetteer shows the distribution of the thirty one surviving mills, plus those sites where the existence of a windmill has been confirmed but where there are no physical remains.

AMLWCH

Melin Adda
(Pentrefelin)
Amlwch

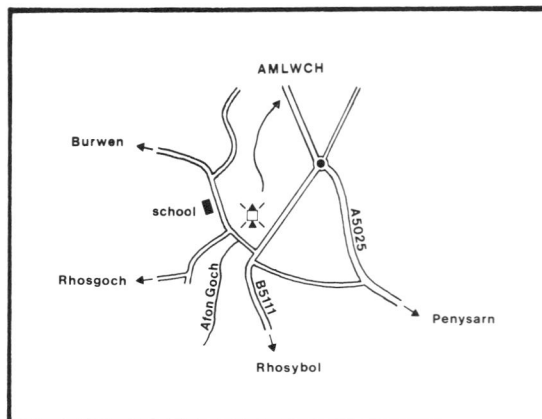

SH 440921

The converted tower of Melin Adda stands on the southern side of Amlwch, next to the car park of Amlwch Leisure Centre and opposite Sir Thomas Jones Secondary School.

Believed to date from the 1790s, Melin Adda was one of three mills in the vicinity, the others being watermills of much earlier origin. One of these, confusingly also called Melin Adda, is mentioned in the 1352 *Extent of Anglesey*. This duplication of names has led to problems differentiating between the two types of mill, particularly with regard to their owners and millers.

However, by the middle of the nineteenth century, both windmill and watermill seemed to be under the same ownership, Pigot's Directory of 1849 and Slater's Directory of 1850 containing the identical entries 'Lewis & Owens, Melin Adda Mills'. It is unlikely that either Lewis or Owens were millers, as Owen Hughes is known to have been running the windmill at that time (and possibly the watermill as well). He was killed in May 1851 when struck on the head by one of Melin Adda's sails — a not uncommon accident. Hugh Hughes (his son?) succeeded him as miller and remained at Melin Adda until his own death in 1865. He was followed by John Williams who was still there in 1889, according to Sutton's Directory of that year.

During this period the windmill also changed owners, an Amlwch trade directory of 1881 recording Melin Adda, Pentrefelin as belonging to 'Messrs. Wm Jones and Son, Corn and Flour Merchants'. **Sutton's Directory** of 1889, which named John Williams as miller, also mentioned John Jones as being at Pentrefelin Adda. He was probably the mill's owner, and may well have been the son of the aforementioned Wm Jones.

Melin Adda closed around 1912 and is marked 'Old Windmill' on the 1918 one-inch O.S. map. By 1929 it was an empty shell and over the ensuing years became increasingly derelict until rescued in the mid-1970s and turned into a residence.

Plate 28
Melin Adda with Amlwch in the distance c.1900.

Melin Adda today.

AMLWCH

Melin y Borth
(Mona Mill)
Amlwch Port

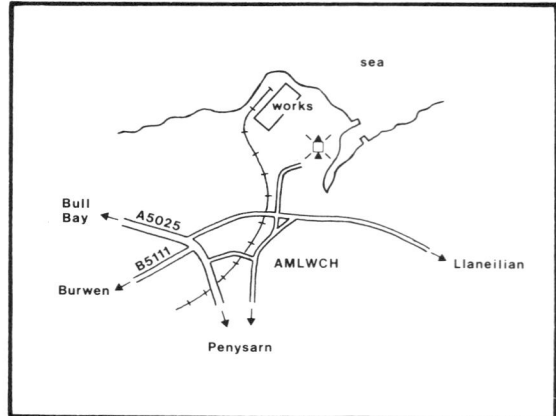

SH 448935

The imposing brick and stone tower of Melin y Borth overlooks the sheltered harbour of Amlwch Port, about half a mile north-east of Amlwch town centre.

With seven floors and standing over sixty feet high it had the distinction of being the tallest windmill to have been built on Anglesey. On the mill's completion in 1816 the **North Wales Gazette** informed its readers that 'A windmill on an improved principle has been erected at Amlwch which will grind 70 bushels of corn in an hour and is considered to be one of the most complete pieces of machinery in the county. Messrs Paynters of Amlwch erected this mill'.

The Paynters were a prominent family in Amlwch during the nineteenth century with wide-ranging business interests. As well as being corn merchants they had connections with the legal profession, insurance, shipbuilding and the timber trade. Although Melin y Borth belonged to the Paynters its day-to-day running was, for many years, in the hands of the Jones family. **Slater's Directory** of 1850 lists John Wynne Paynter of Maesllwyn as corn merchant and William Jones of Queen Street as miller, the latter probably being the son of one Robert Jones who died in 1861 and who is described on his headstone in Amlwch churchyard as 'Miller, Amlwch Port'. William Jones continues to appear as miller of Melin y Borth in various Directories until 1895, not long before its closure.

During his tenure a tragic accident occurred at the mill. In its edition on 6th November a magazine called **The Miller** reported that 'William Jones, a youth aged 18, son of Mr. Jones, proprietor of Amlwch Port Windmill, whilst at work inside the mill on Tuesday, was struck by lightning and killed.'

Although Melin y Borth now stands alone, on waste ground between council houses and the old Shell oil terminal, it was once at the centre of a small cluster of buildings. Photographs taken at the turn of the century show several huddling together in the shadow of the mill's wide gallery. This encircled the tower at second floor level, being supported on wooden piers and, for part of its circumference, by the roof of one of the buildings.

Plate 29
Melin y Borth, Amlwch Port c.1900.

The mill's capacious interior — its basement's internal diameter measured some thirty feet — not only provided ample storage space but also enabled four pairs of millstones to be accommodated. Each pair came to have a specific use; one ground wheat, one barley, one oats and one Indian corn (maize).

Melin y Borth is thought to have ceased working in the early years of this century since when its condition has steadily deteriorated. The sails and machinery were removed, probably for scrap, and the surrounding buildings demolished. Today it stands in splendid isolation, a gutted tower with only part of the cap frame timbers remaining. Years of neglect and vandalism have taken their toll, leading to concern about the mill's safety. Two iron bands have been bolted around the tower to strengthen it and a wire fence erected to deny access to local children. Following the closure of the oil terminal, on whose land it stood, Melin y Borth has become the responsibility of Anglesey Borough Council. Its future remains in doubt; demolition is one option being considered, full restoration is out of the question. Hopefully, steps will be taken to make the structure safe and preserve it as a landmark, perhaps with an internal staircase ascending to a viewing platform.

Plate 30
The crumbling tower of Melin y Pant.

AMLWCH

Melin y Pant,
Porth Llechog
(Bull Bay)

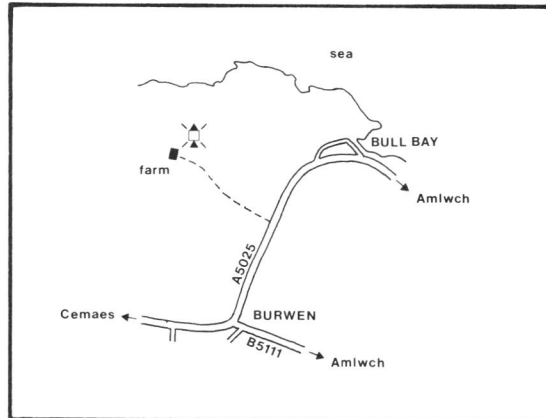

SH 416 943

The ruined remains of this small tower mill belong to Pant y Gaseg farm which lies at the end of a long track leaving the A 5025 between Bull Bay and Burwen. The sea is only a quarter of a mile away and on most days strong winds sweep in over the heather-clad cliffs.

Nothing is known of the mill's origin or history, but its small size suggests that it functioned as a domestic farm mill similar to those at Treban-Meirig and Fferam (see page 00). Indeed, the tower's internal diameter of less than ten feet makes it extremely unlikely that it ever contained millstones for the purpose of grinding corn. More probable is that the power provided by the sails was diverted into the adjacent barn to operate machinery for chaffing hay, chopping roots and churning butter. Although one side of the tower has collapsed (and been partly rebuilt) it is still possible to distinguish the outline of an opening near the base through which a shaft or belt drive could have passed.

Improvements in farm machinery in the late nineteenth century and the availability of more reliable sources of power spelt the end for farm mills like Melin y Pant. It has been a ruin for as long as anyone can remember and was already disused when the Ordnance Survey mapped the area in 1899.

BODFFORDD

Melin Frogwy,
Bodffordd

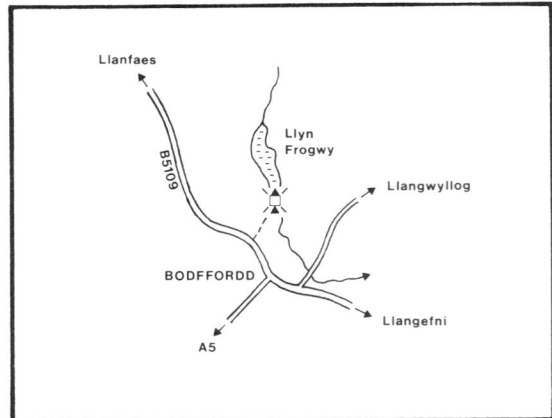

SH 426773

The ivy-clad tower of Melin Frogwy lies at the end of a wooded lane which branches from the B5109 just west of the village at Bodffordd. It occupies a rocky outcrop at the southern edge of Llyn Frogwy, a small lake which once supplied water to the overshot waterwheel of an adjacent watermill.

Whereas the windmill is probably only early nineteenth century in origin there has been a watermill on the site for over 600 years. In the *Extent of Anglesey* it is recorded that 'Llywelyn ap Dafydd Fychan . . . has his own mill called Melin Bodffordd'.

The watermill is also mentioned in the diary of William Bulkeley. On April 24th 1735 Bulkeley, in his legal capacity, attended the court at Beaumaris where the miller of Frogwy, John Davis, had been indicted for killing his maid. Found guilty of manslaughter Davis was sentenced to be burnt in the hand, a punishment which reveals much about the low status of maidservants in the mid-eighteenth century! No details were given as to the severity of the burning, nor whether it affected Davis' ability to work the mill.

Quite why a windmill should have been needed so close to such a successful watermill is puzzling. Although Anglesey had experienced several periods of drought, Llyn Frogwy would have provided the watermill with one of the most reliable sources of supply on the island. Whatever the reason, a windmill was built, but it does not seem to have had a long working life and by the beginning of the 1890s had fallen into disuse. William Roberts, who farmed at nearby Gafrogwy Fawr and owned both the mills, decided to convert the windmill into a dwelling, adding a single storey extension to serve as a kitchen.

Retired millwright Glyn Rowlands spent his childhood in the 'mill house' and vividly remembers the buffeting it received from the strong south-westerly winds channelled down the valley by the contours of the surrounding hills. His father was miller of the watermill and the family lived in the converted tower until 1948 when the watermill closed. As a boy Glyn Rowlands helped his father in the

66

Plate 31
Frogwy's converted windmill and working watermill at the beginning of this century.

Today the watermill is a residence while the windmill tower serves as a store.

mill, gradually acquiring his skills as a millwright. Rowlands senior was also a grain merchant, but because the amount of locally grown grain was insufficient to sustain his business additional supplies had to be brought in from outside Anglesey. Glyn Rowlands particularly recalls high quality Indian corn being imported from Rumania in the 1930s. Small quantities of linseed also passed through the mill, not for oil as was the case elsewhere, but to produce a very nutritious animal food much sought after by local farmers who reserved it for those calves they considered potential champions at agricultural shows.

In the photograph a line of poles can be seen climbing the hill from the watermill to Gafrogwy Fawr on the skyline. The poles formed part of an ingenious device which utilised the energy generated by Frogwy's powerful waterwheel to drive agricultural machinery. A long shaft bevelled to the watermill's pit wheel transmitted power to a revolving drum around which passed an endless cable. This moving cable, supported on runners attached to the poles, travelled to the farm some quarter of a mile away where it passed round a similar drum before returning downhill. A belt drive taken from the drum could be engaged to work machines for chaffing hay, chopping roots, etc. Although successful the equipment was dismantled following a tragic accident on 2nd June 1909 when Hugh Hughes the miller, bending over to oil the drum's axle, caught his loose apron in the moving cable and was dragged into the machinery and killed. He was only 44 and left a wife and five children.

Plate 33 Melin Manaw.

BODFFORDD

Melin Manaw

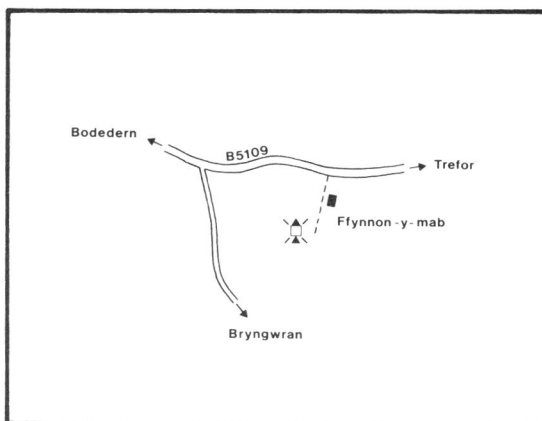

SH 359795

The semi-circular remains of Melin Manaw stand on a grassy rise at the end of a private track which threads its way between the buildings of Fynnon- y- Mab farm. The track runs south from the B5109 approximately three quarters of a mile west of Trefor crossroads.

As with many of Anglesey's windmills few facts have come to light concerning the history of Melin Manaw. It was already in a disused state by the First World War — so much so that when the farm on which it stood came up for auction in September 1919 the mill received only a brief mention. The farm, also known as Melin Manaw, was advertised as 'An attractive Smallholding . . . in the occupation of Mr Thomas Hughes as yearly tenant at a rental of £15'. Although the mill is clearly shown on the plan of the property no reference to it appears in the written description, indicating that by then it had deteriorated to such an extent as to be of little interest to a potential buyer.

Today the structure is in a parlous state. One side has totally collapsed and elder trees push out from cracks in the lichen-encrusted stonework. A doorway and two window apertures survive in that part of the tower which still retains its original height. Given its present condition and the unlikelihood of any repair work being carried out, Melin Manaw seems destined soon to join Anglesey's growing number of lost windmills.

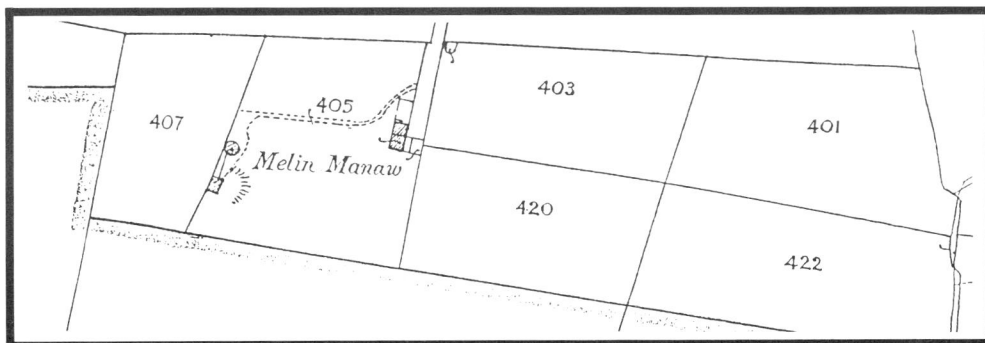

BODFFORDD

Melin Newydd
Tre'rddol

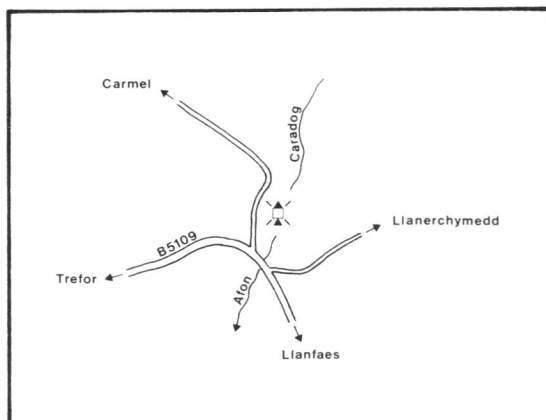

SH 390803

   Not far from Melin Manaw is the well-preserved tower of Melin Newydd, Tre'rddol, at the side of the minor road which winds north from the B5109, one mile east of Trefor crossroads.

   Built on land belonging to the Bodorgan Estate, Melin Newydd was one of the last windmills to be erected on Anglesey, the date 1833 being inscribed above its north-facing doorway, together with the initials of Owen John Augustus Fuller Meyrick.

By his marriage to Clara Meyrick, Owen Fuller had succeeded to the Bodorgan Estate in 1825, adding the family name Meyrick to his own in accordance with his father-in-law's wishes. With his new status as the third largest landowner on the island he stood for Parliament in 1837 as the Tory candidate in the Anglesey by-election, but was defeated by William Owen Stanley of Penrhos, the Whig representative. Disillusioned, Meyrick withdrew from the General Election later that year leaving William Stanley to be returned unopposed.

The name Melin Newydd, or New Mill, was given to distinguish it from an existing watermill some hundred yards away (of which nothing now remains). As both mills were under the same ownership they could be worked in conjunction, a practice not uncommon elsewhere on Anglesey. Having two different mills meant the miller could choose which would be the more suitable on any given occasion, usually opting for the windmill on windy days and during times of drought.

Sutton's Directory names Rowland Roberts as miller of Melin Newydd during the 1890s, and it is not unreasonable to assume that previous generations of the same family had been at the mill before him. He was followed by William Roberts who worked Melin Newydd until the early 1920s when it sustained serious damage in a storm and was forced to close. By the end of the decade one sail had gone and the skeletal ribs of the other three hung from a cap which

Plate 32
Melin Newydd in 1929. The cap and sails have since gone but much of the stone platform still survives.

71

had developed a pronounced lean. Eventually, the decaying and dangerous remains were taken down and the building emptied of its machinery.

The tower today is still structurally sound with two of its internal floors remaining in place, a result of the protection afforded by the corrugated iron roof fitted after removal of the cap. It is currently used as a shippon, the cattle being kept company by large numbers of roosting pigeons.

Unusually, Melin Newydd has managed to retain a substantial part of its encircling stone platform, although a section has been dug away by the present owner to allow access for farm machinery.

Plate 34
Melin Hermon c.1905. The adjacent cottage is now derelict.

BODORGAN

Melin Hermon
(Tyddyn Olifer)

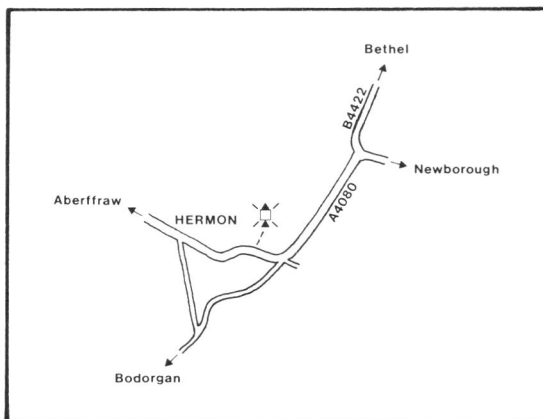

SH 390690

For travellers on the A4080 between Aberffraw and Newborough the capless tower of Melin Hermon is a noticeable landmark on the skyline. Standing on a ridge close to the small village of Hermon, the mill is reached by a narrow lane which leaves the main road a few yards west of the chapel.

According to William Bulkeley, construction of the mill began in the late spring of 1743. On May 8th of that year he wrote in his diary: 'Foundation of the Bodorgan Mill was laid this day' — so called because it was on the Bodorgan Estate of the Meyrick family. Like most large landowners on Anglesey the Meyricks owned a number of mills, although at that time Melin Hermon was their only windmill. The others consisted of two fulling mills (Pandy) and six water powered corn mills, including one on the shore at Llanfair-yn-Neubwll described as being "watered by the tide".

Corn mills were generally more valuable than fulling mills and this was reflected in the rents charged. In 1774 the rent for Melin Hermon and twenty acres of land was £20, almost three times the £7.7s. charged for Pont y Pandy near Llanddeusant which had a similar acreage. Using this criterion, Llanfair-yn-Neubwll tide mill appears to have been the most valuable property, a rent of £20 being charged for the mill and eleven acres of land (the same rent as for Melin Hermon, but with little more than half the acreage).

Over the ensuing years the Bodorgan Estate gradually acquired more mills, one of which was a watermill on the Afon Gwna not far from Melin Hermon. Melin Gwna was built towards the middle of the nineteenth century by John Griffith, on land originally leased from the Hughes family of Llys Dulas. When John Griffith died the tenancy of the watermill passed to his son, also called John Griffith. He bought Melin Hermon from the Meyricks sometime in the 1880s, probably in order to reduce the competition rather than increase his workload.

The second John Griffith did not marry until late in life and on his death in

1909 left a young wife and four children. The eldest son — yet another John Griffith — eventually took over the running of Melin Gwna, having also inherited Melin Hermon which, by that time, had stopped working. However, he really wanted the family smallholding near Malltraeth which had been left to his two brothers, Hugh and Gwilym. This he managed to obtain by paying a sum of money to Hugh and transferring the ownership of Melin Hermon to Gwilym.

Sadly, Gwilym fell ill and was no longer able to look after his own affairs. Brothers John and Hugh became trustees of Melin Hermon and let the mill to a former employee who fitted a corrugated iron roof and used the tower to store hay. When Gwilym Jones died in 1989 the family decided to sell the mill and Melin Hermon is now owned by Owen Thomas, landlord of the Royal Oak Inn at Malltraeth.

Although Melin Hermon's working life had ended before the First World War, Melin Gwna continued in operation until the early 1950s, mainly grinding animal feed for local farmers. It was finally forced to close when the wooden leat supplying water to the millwheel collapsed and the mill's owner refused to pay for repairs. In 1978 there was talk of restoring the mill as a museum, but nothing came of the idea and Melin Gwna was left to become derelict.

Less than a mile away Melin Hermon still looks out over Llyn Coron and the dunes of Aberffraw to distant Holyhead Mountain. Except for a few rotten timbers the roofless tower is now empty, serving only as a vantage point for perching jackdaws.

CWM CADNANT

Melin Llandegfan

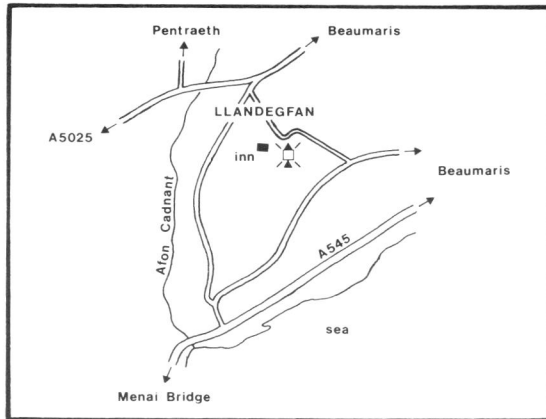

SH 567740

Melin Llandegfan is situated on the north-west edge of Llandegfan village near the Pen y Cefn Inn, on a sharp bend in the minor road which runs from the parish church towards the coast. However, it can easily be missed as the tower is so smothered with ivy that its original shape is not immediately discernable.

Plate 35
Ivy has almost completely covered the old tower of Melin Llandegfan.

The mill probably dates from the 1820s, but was certainly in existence by 1831 as it appears on Dawson's map of that year. It is thought to have ceased working sometime before the First World War, and is marked 'Old Windmill' on the 1918 one-inch O.S. map. According to Rex Wailes the cap and remains of the sails were still in place in 1929, but by 1937 they had disappeared — as had several courses of stonework from the top of the tower, reducing its original five storeys to nearer four.

Not long afterwards Melin Llandegfan began its second career as a water tower. The local Water Authority considered that its elevated site would be ideal for a reservoir to supply Llandegfan village and, to this effect, installed a large storage tank inside the empty tower. Water was abstracted from the nearby Afon Cadnant, filtered and pumped uphill into the tank. The scheme was initially successful, but during the 1950s and 1960s Llandegfan experienced considerable residential development and the volume of water contained in Melin Llandegfan's tank became inadequate to satisfy the rapidly rising demand. A new source of supply became essential and this was provided by the recently completed Cefni Reservoir near Llangefni. Water was piped directly to the village and Melin Llandegfan once again became redundant.

Although the pipework to the tank was disconnected over twenty years ago the tank itself remained within the tower and was not finally removed until 1989, following a request from the mill's owner.

Melin Llandegfan's future is uncertain as it becomes increasingly overgrown by ivy. This now has such a hold, penetrating deep into the stonework, that any attempt to remove it could result in serious damage to the tower's structure.

CYLCH Y GARN

Melin Drylliau
(Crugmor Fawr),
Church Bay

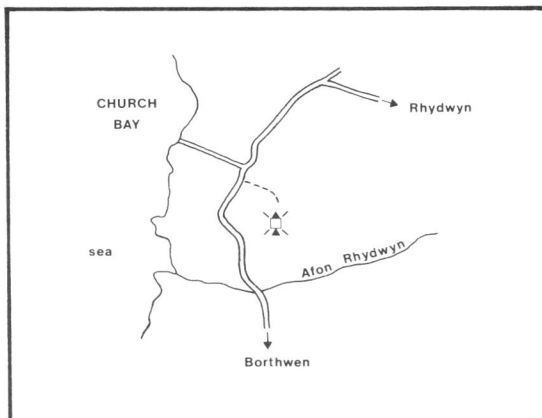

SH 305887

Visitors to Church Bay cannot fail to notice the disused tower of Melin Drylliau, conspicuous on the skyline above the scattered houses. The mill stands at the head of a short track which climbs steeply from the minor road between Church Bay and Borthwen.

Clearly visible from Holyhead Bay, Melin Drylliau has long been a prominent landmark for sailors. It is shown on Captain Beechey's sea chart of 1840 and also on the 1881 Admiralty chart, being named Rhyddlad Mill on the latter. Melin Drylliau probably dates from the early nineteenth century and was one

Plate 36
Melin Drylliau and its adjacent granary.

of two windmills in the locality, the other (now demolished) standing half a mile further inland at Rhydwyn.

For much of the last quarter of the nineteenth century Melin Drylliau was in the hands of John Rowlands, one of the famous family of Anglesey millers. He was succeeded by his son, Rowland William Rowlands (known locally as Rowland Williams), who worked the mill until it was destroyed by fire in 1914. Kitty Williams, now nearly ninety but then a young girl, remembers running to the mill on hearing it was alight, arriving just in time to witness the blazing cap and sails come crashing to the ground.

The loss of Melin Drylliau was the second tragedy to befall Rowland William Rowlands, his young son having been killed a few years earlier. Playing too near the mill he had been struck by a moving sail as the cap was being turned into the wind. Despite these setbacks Rowland William Rowlands continued in business on the site for a number of years. Operating from the granary next to the burnt out mill he traded in animal feed which he bought wholesale from Holyhead and retailed to farmers in the neighbourhood.

Since the fire the gutted tower has stood empty, structurally still in good condition, but of little practical use.

HOLYHEAD

Melin yr Ogof
(George's Mill)
Kingsland

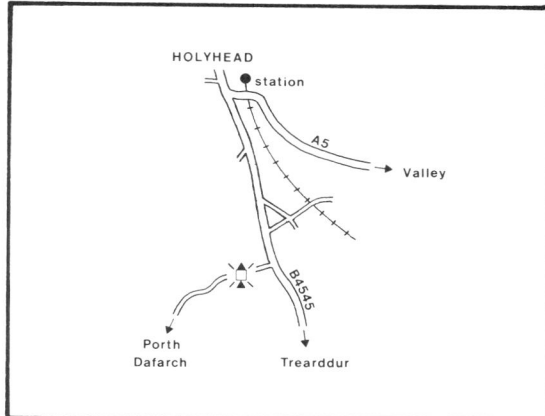

SH 248811

The white painted tower of Melin yr Ogof, Kingsland, stands on high ground overlooking Holyhead Leisure Centre. It is located just off the B4545 to Trearddur Bay, along the appropriately named Mill Road.

The mill's elevated position on the southern outskirts of Holyhead made it an important navigational aid for ships entering and leaving the port, and it appears as such on a number of nineteenth century sea charts. Melin yr Ogof can be seen to the right of Skinner's Monument on the sectional view which accompanies Captain Beechey's chart of 1840.

Thought to date from around 1825, Melin yr Ogof was built by Hugh Hughes of Ty Mawr farm on land belonging to the Stanley family. By the 1850s relations between Hughes and his landlord had become strained following rent increases. Writing in March 1858 to his son William, who had emigrated to Iowa, a disenchanted Hughes complained that 'Mr Stanley has made Great Allterations with all his farm Tenants, he has Raised the Rents on us all . . . he has Taken the Land from the Mill & Let it to Hugh Roger now I have no Place to turn the Horses if they are sick or Lame but to the Road, and the Mill will not Let for so much a year by 6£ that is a lost 120£ interest with me William after building on his Honour, Very Poor Honour . . . I would never built the Mill if I knew he would take the Land.'

Plate 37
Melin yr Ogof c.1920.

Plate 38 The Sack Hoist Mechanism. Carved on the wooden panelling around the upright shaft are what appear to be patterns for dressing the millstones.

Plate 39 Stone Nut and Great Spur Wheel. The wooden lever is moved to engage or disengage the drive to the millstones.

Plate 40 The Stone Floor with its three pairs of millstones.

Plate 41 A Runner Stone exposed by the removal of a section of its polygonal wooden casing.

Plate 42 Governors and Tentering Gear.

Plate 43 Flour Dresser.

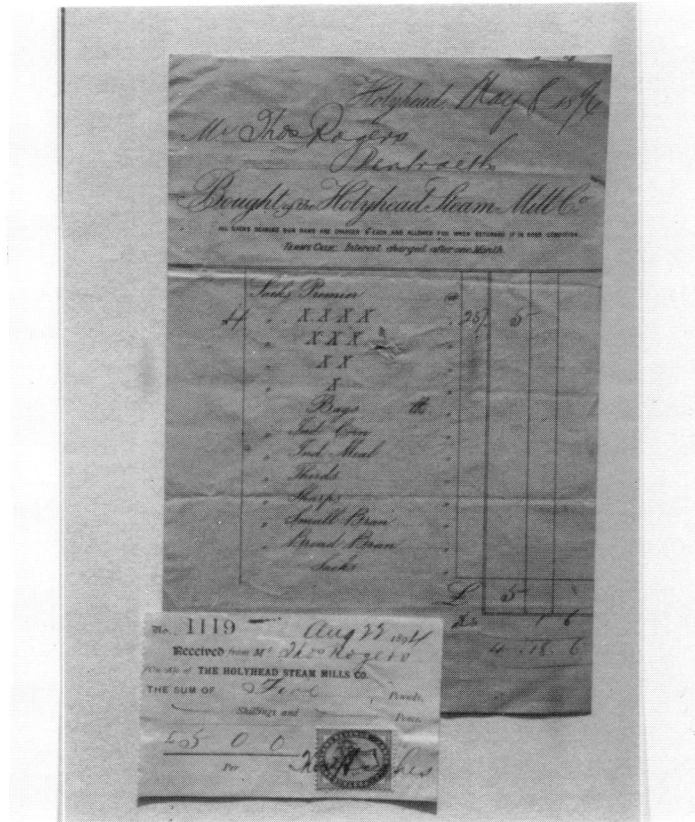

Holyhead Steam Mill closed in 1907, its machinery being adapted to help generate electricity for the town's houses and street lights.

Hugh Hughes died in 1869, bequeathing the mill to his widowed daughter Margaret and her sister Mary. This was not the best of times to inherit a windmill, a new steam-powered mill having recently opened in Kings Road, Holyhead. However, much of the steam mill's business involved the processing of foreign grain shipped into the port from Liverpool, the flour produced then being bagged and distributed by cart throughout Anglesey. The invoice suggests that it was worthwhile for Thomas Rogers of Pentraeth to purchase flour from the steam mill twenty miles away rather than buy from more local sources. However, he must have been tardy settling the bill because he appears to have forfeited his 1*s*.6*d*. discount!

Although the opening of Holyhead's steam mill helped bring about the closure of nearby Melin Ucheldre, Melin yr Ogof managed to survive the competition, partly because it was further from the steam mill and partly because it offered a service that some farmers still wanted, namely to have their corn ground in exchange for payment in kind. When Rex Wailes visited Melin yr Ogof in 1929 he remarked on the toll cupboards or *kests* which were used for such payments. By then the mill had not worked for nearly ten years, having been forced to close due to a crack in the stone neck bearing which supported the windshaft. Landlord and tenant could not agree who should stand the cost of renewal and Melin yr Ogof has not worked since.

The cap and bare ribs of three of the sails remained in place until 1939 when a storm tilted them forward. Because the cap was potentially dangerous (and to prevent the mill being used as a bearing for enemy shipping in the Irish Sea) it was removed by George Lock and Sons, scrap merchants of Porth Penrhyn, Bangor. Attempts to dislodge the windshaft with gelignite failed and it had to be pushed off using a block and tackle, eventually overbalancing and toppling into the adjoining field. The rest of the cap was subsequently removed with explosives, resulting in some debris falling near Kingsland's Ebenezer Chapel nearly 200 yards away.

Life at Melin yr Ogof before its closure was recalled by Hugh Ross in the Summer 1961 edition of the magazine **Môn**. As a boy he was a frequent visitor to the mill and remembered the carts arriving full of corn and being allowed to stay to watch the millstones at work. One particular incident stuck in his memory concerning rules and superstitions in the mill: 'I did not know at that time but someone was called by his nickname (considered to bring bad luck) and almost at once everyone took off his cap and one fetched a stool and put it in the middle of the floor. Someone else fetched a plank about three feet long and six inches wide with a handle, and everyone else stood in a circle trying to decide the punishment. Eventually two got hold of him, put him face down over the stool and gave him three strokes on his backside.'

Hugh Ross also remembered the miller, William Owen, confessing to having nightmares about not being able to stop the mill because the brake had failed. He was probably not the only miller to be kept awake by this worry!

After the cap and sails were removed in 1939 the top of the tower was concreted over, enabling the machinery inside to survive in good condition until the present day (see Plates 38 to 43). Melin yr Ogof is now the only windmill in Wales with its original machinery still in place, a fact which has prompted the Welsh Mills Society and other concerned parties to campaign for its preservation.

Plate 44
John Rowlands outside Melin Cemaes c.1910. To his left is the miller's cottage and its adjoining granary, the lower floor of which served as the dairy. A cowshed completed the line of buildings but is not included in this view.

LLANBADRIG

Melin Cemaes

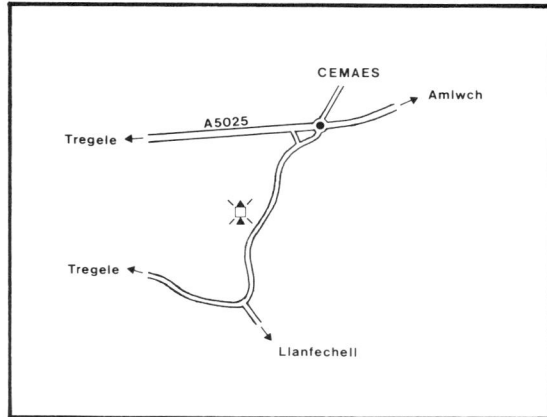

SH 365925

Melin Cemaes stands in a prominent position three quarters of a mile south-west of Cemaes, at the side of the minor road which runs from the harbour to Llanfechell village.

During the 1820s coastal trade between Cemaes and Liverpool had been steadily growing, particularly in agricultural produce, and in order to consolidate this growth an agreement was reached in 1828 to improve port facilities at Cemaes, the work to be financed by contributions from local people and a number of Liverpool businessmen.

It was in anticipation of increased trade with Liverpool that the windmill at Cemaes was built in 1828, on land belonging to Hugh Williams. According to Llanbadrig parish register John Jones was the miller in 1836, but by 1851 the mill had changed hands, the Census for that year and 1861 naming John Williams as miller. He was followed by William Rowlands, brother of the miller of Melin Drylliau Church Bay. **Sutton's Directory** confirms that William Rowlands was still at Melin Cemaes in 1895, and it is likely that he continued there until the early years of this century before handing over the running of the mill to his nephew John Rowlands.

In 1918 Melin Cemaes and its five acres, two roods and seven perches of land were purchased by John Richard Roberts, the son of Isaac Roberts a well-known Anglesey millwright. The Reverend John Rowlands of Llanfachraeth recalls his father, the miller at Melin Machreth, being full of praise for Isaac Roberts' ability as a millwright: 'My father would meet him at Fferam Gyd, Llanbabo, with a pony and trap and bring him to Llanfachraeth. He would sit down, light his pipe, survey the mill for a time, and then he knew exactly what had to be done.'

John Roberts, on the other hand, had the reputation of being rather eccentric and, according to his daughter-in-law Connie Jones, even his wife and children found him difficult to live with. A devout Baptist who cycled to church

87

Plate 45
Melin Cemaes undergoing re-roofing in 1939. The Austin Seven belonged to Emyr Roberts, the miller's son.

three times every Sunday he became very successful as a singer, winning prizes at several eisteddfodau. However, as the owner of a windmill in the years immediately following the First World War he was faced with a different sort of competition. By the late 1920s Melin Cemaes was one of only a handful of windmills still at work on Anglesey and John Roberts was fighting a losing economic battle. Although Rex Wailes' survey of 1929 found the mill complete with 'cap and four sails' its days relying on wind power had all but ended. Efforts to prolong its working life by installing a diesel engine were successful for a time, enabling grinding to continue throughout the Second World War, but in 1946 Melin Cemaes finally bowed to the inevitable and ceased operations.

By then its sails had gone, although fortunately the cap remained enabling the mill to survive in a reasonable state of repair for a number of years. Its relatively good condition, plus the fact that the machinery inside was intact and in position, resulted in Melin Cemaes being considered as a possible candidate for restoration by Anglesey County Council in 1954.

Sadly, a detailed survey revealed that the mill would prove very costly to restore and it was rejected in favour of Melin Llynon, Llanddeusant. Following this decision Melin Cemaes began to deteriorate; the machinery went for scrap, the roof decayed and the floors gave way. By the late 1970s the mill was derelict.

Its owner applied for and was granted planning permission to convert the tower into a dwelling. After several years of intermittent building work Melin Cemaes was put up for sale, the estate agent's description offering potential buyers a 'rare opportunity to purchase recently renovated mill with panoramic views to the coast'. Such views could be obtained from the large picture window set into the top of the tower, one of a number of alterations to the original structure which have not resulted, purists would argue, in the most sympathetic of conversions. Another, the addition of a steeply pitched roof with central chimney, has earned Melin Cemaes the unflattering nickname 'Noddy's house' among locals.

Melin Cemaes today.

LLANDDYFNAN

Melin Llanddyfnan
(Pen y fan)

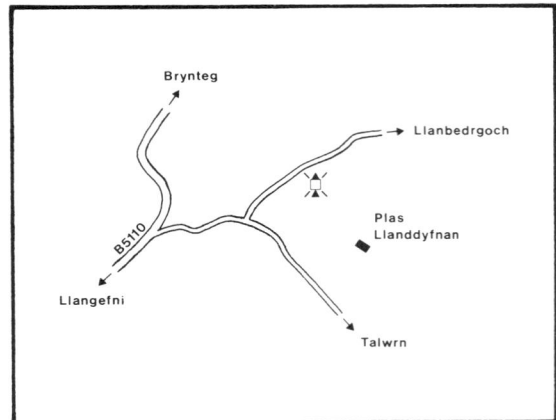

SH 484786

Melin Llanddyfnan can be seen beside the minor road to Llanbedrgoch which branches from the B5110 two miles north-east of Llangefni. It stands on a rise overlooking the house and grounds of Plas Llanddyfnan.

The earliest record of the mill's existence is on an estate map of 1746, and this points to its being part of the upsurge in windmill construction which took place on Anglesey during the late 1730s and early 1740s, a time when the island's watermills were suffering from recurring droughts. Unfortunately, as is the case with many of Anglesey's windmills, Melin Llanddyfnan's history has

Plate 46
Melin Llanddyfnan and granary around the turn of the century.

90

gone largely unrecorded and only two of its millers have come to light, namely John Gray in 1850 and Robert Owen in 1883. The mill is known to have been working around the turn of the century, but seems to have closed soon afterwards and is marked 'Old Windmill' on the 1918 one-inch O.S. map.

Melin Llanddyfnan stood roofless and empty until the 1950s when its then owner, Francis Wilson Q.C. (the Recorder of Chester), decided to restore it, principally for use as a viewing platform. Although the approach to the mill from the road is by a gentle, wooded slope, the ground to the south and east falls away sharply enabling extensive views to be obtained from the top of the tower. The restoration work consisted of replastering the interior walls, fitting new floors, stairways and windows, and adding a flat roof with trap- door access. Electricity was also installed.

Around the same time, the old granary next to the mill was turned into a dwelling by the mother-in-law of the mill's present owner Mr M.J.S. Preece, the conversion earning her second prize in an Ideal Homes competition.

The mill itself now serves as a store and is in need of some attention; there are visible signs of damp and several floorboards appear unsafe.

LLANDDYFNAN

Melin Llidiart
Capel Coch

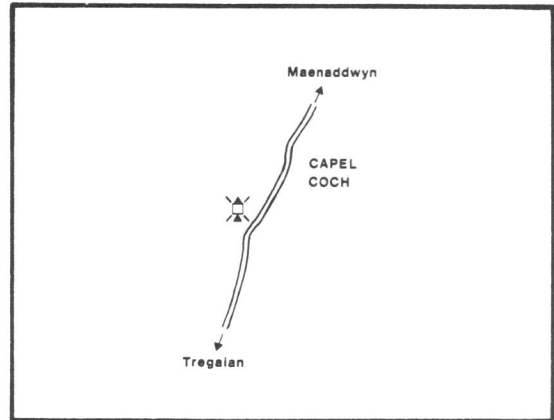

SH 457820

   Probably one of the oldest surviving windmills on Anglesey, Melin Llidiart stands at the southern edge of Capel Coch, a straggling hamlet on the minor road from Tregaian to Maenaddwyn.
   The mill is thought to date from around the middle of the eighteenth century, although nothing seems to be known of its early history. However, by 1883, according to **Slater's Directory**, Melin Llidiart was being worked by Hugh Pritchard of Brynfelin. He must have died by 1895 because that year's edition of the Directory names his wife Jane as miller. Not long afterwards the mill

Plate 47
Melin Llidiart's capless tower can be seen to the right of the people standing outside the blacksmith's shop in Capel Coch's main street. The photograph dates from the early years of this century.

was irreparably damaged in a severe storm, losing its cap and sails. It never worked again.

For many years Melin Llidiart stood abandoned and overgrown, difficult to reach through brambles and nettles which had been allowed to run wild. Recently, a change of ownership has resulted in the ground around the mill being cleared, revealing once more the sharply tapering, red sandstone tower. The new owner hopes to develop the site, but in April 1991 his application for planning permission to convert the old house next to the mill into a four bedroom hotel and erect six holiday cottages was turned down by the Planning Committee of Anglesey Borough Council.

Plate 48
The converted tower of Melin Gallt y Benddu now provides holiday accommodation.

LLANERCHYMEDD

Melin Gallt y Benddu

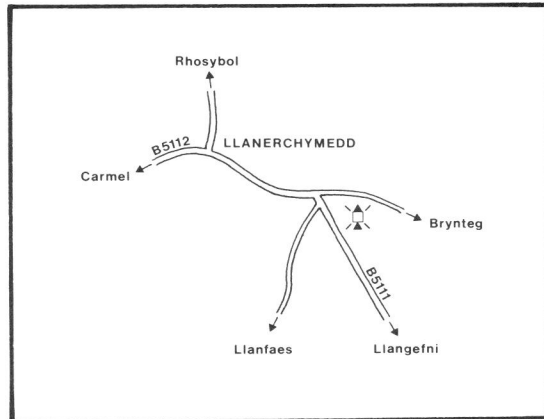

SH 426838

From the centre of Llanerchymedd the B5111 climbs a long hill towards the white painted tower of Melin Gallt y Benddu, located between the 'B' road and the minor road to Brynteg.

The diary of William Bulkely records that construction of the mill began on September 8th 1737 and was completed on October 24th of the following year, making Melin Gallt y Benddu the oldest known tower mill still in existence on Anglesey. Thomas Davis was the miller in 1745, but whether he had been at the mill since its opening is not known. He paid a land tax for the mill of one shilling, as did Evan Ellis who succeeded him, but by 1763, when John Williams was running Melin Gallt y Benddu, the tax had doubled to two shillings.

Melin Gallt y Benddu's history is then poorly documented until September 1847 when the mill was advertised for sale as part of the Llwydiarth Estate:

> 'The Windmill is well built, is worked with four pairs of
> stones, has excellent machinery and is in good repair.'

The Census four years later revealed the miller to be another John Williams, described as a 'master miller employing six men'. It is not inconceivable that he was related to his namesake from the previous century. Already sixty four years old in 1851, John Williams was still at Melin Gallt y Benddu at the time of the next Census, but is thought to have retired soon afterwards. He lived until 1873, outlasting his wife by nearly forty years.

Melin Gallt y Benddu then passed into the hands of John Jones of Melin Rhos Fawr, Brynteg. Keen to see his sons follow in his footsteps he installed Edward, the eldest, as miller of Melin Gallt y Benddu. Edward worked the mill until his marriage when he left to take over Abergwyngregyn mill on the mainland. His brother John replaced him at Melin Gallt y Benddu and is named as miller in *Worrall's Directory* of 1874. However, John Jones was a less than enthusiastic miller and eventually gave up the business to train as a preacher, leaving younger brother Hugh in charge. Only fifteen years of age in 1881 Hugh Jones was very young to be responsible for a windmill but, nevertheless, managed successfully until the late 1880s when he married and became a shopkeeper, so ending the family connection with Melin Gallt y Benddu.

Owen Roberts then took over Melin Gallt y Benddu and was the last person to work the mill, a violent storm removing its cap and sails just before the turn of the century. During the twentieth century Melin Gallt y Benddu slowly deteriorated before being bought in 1964 by Edgar Williams. He converted the empty tower into a dwelling which is now let as holiday accommodation.

LLANFAELOG

Melin Maelgwyn
(Melin Uchaf)
Bryndu

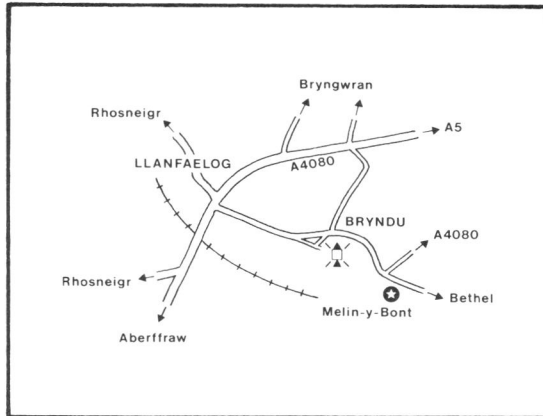

SH 342728

The derelict five storey tower of Melin Maelgwyn stands in a corner of a farmyard just off the minor road from Llanfaelog village to Ty Croes railway station. It is the more westerly of two windmills in the vicinity, the other being Melin y Bont (Melin Isaf), a mill which was also worked by water.

Set in the tower a stone tablet with the initials $_OR_K$ and the date 1789 reveals Melin Maelgwyn to be late eighteenth century in origin, built over thirty years before its near neighbour. It appears to have had a busy but uneventful career, ceasing to work by wind some seventy years ago, although continuing to grind corn for several more years with the aid of a diesel engine.

The mill and the farm of which it is part have belonged to the Lewis family for well over one hundred years. 'John Lewis & Son, Maelgwyn' is a frequent entry in trade directories of the 1880s and 1890s, but whether John Lewis was the first of the family to own the mill is not known. However, this long family connection with Melin Maelgwyn could soon be ended as the present owner has recently put the farmhouse up for sale and the mill may follow.

Melin Maelgwyn itself is now an empty shell, the machinery and all the floors having long since gone. A corrugated iron roof was fitted some time ago to keep the rain out and enable the tower to be used as a hay store.

Sadly, Melin Maelgwyn has been largely ignored in favour of its more illustrious neighbour and relegated by photographers to background appearances in their pictures of Melin y Bont and its mill pond.

Plate 49
Melin Maengwyn in 1929. The cap and skeletal remains of the sails have long since disappeared, as has the substantial stone platform which once encircled the tower.

LLANFAELOG

Melin y Bont
(Melin Isaf)
Bryndu

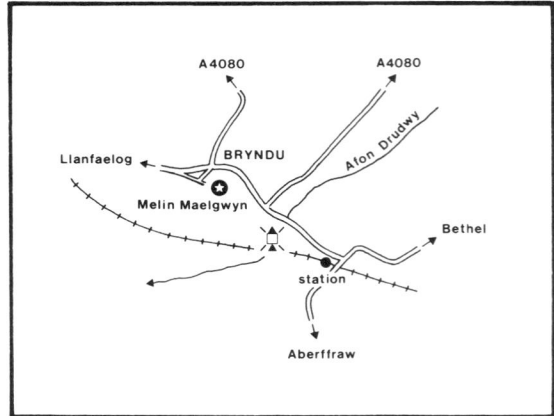

SH 346725

Less than half a mile south-east of Melin Maelgwyn, at the side of a small stream lined with yellow irises, is the abandoned tower of Melin y Bont.

Built in 1825 Melin y Bont uniquely combined both windmill and watermill within the same five-storey structure. An upright shaft running through the mill from top to bottom enabled different pairs of millstones to be driven, by wind from above or by water from below (see diagram). Melin y Bont's position on a slope effectively provided it with two ground floors. The lower of the two housed a 16' diameter breastshot waterwheel (A) supplied by a wooden leat from a pond in front of the mill. The drive from the waterwheel was taken via a cast-iron pitwheel and wallower (B) to a claw clutch on the main vertical shaft located on the 'upper' ground floor. Using a long wooden lever and moveable pivot stand known as a 'monkey' (C) the miller could lift the drive out of engagement, leaving the upright shaft free to revolve. Two floors higher a similar clutch connected with the wind drive; this could also be raised out of gear when required, but here the disengagement was achieved by means of an iron jack fixed to the floor and lifted against a flange on the upper half of the clutch (D). This unusual arrangement of the gearing meant that Melin y Bont's sails revolved in a clockwise direction, the only ones on Anglesey to do so.

Its water power facility gave Melin y Bont a distinct advantage over other mills when there was no wind. Indeed, after the sails were taken down in 1930 John Williams, the last miller, continued to work the mill for several more years using the waterwheel and one pair of stones.

As well as grinding corn the mill machinery was also utilised to perform other tasks. A belt drive taken through a wall aperture turned butter churns in an adjoining building, while another operated a grindstone for sharpening tools. More remarkable was an iron shaft which led from the main drive to a connecting flange on the exterior wall of the mill. From it a series of rods ran in a stone covered trench to farm buildings some fifty yards away, transmitting power

98

MELIN Y BONT, BRYNDU

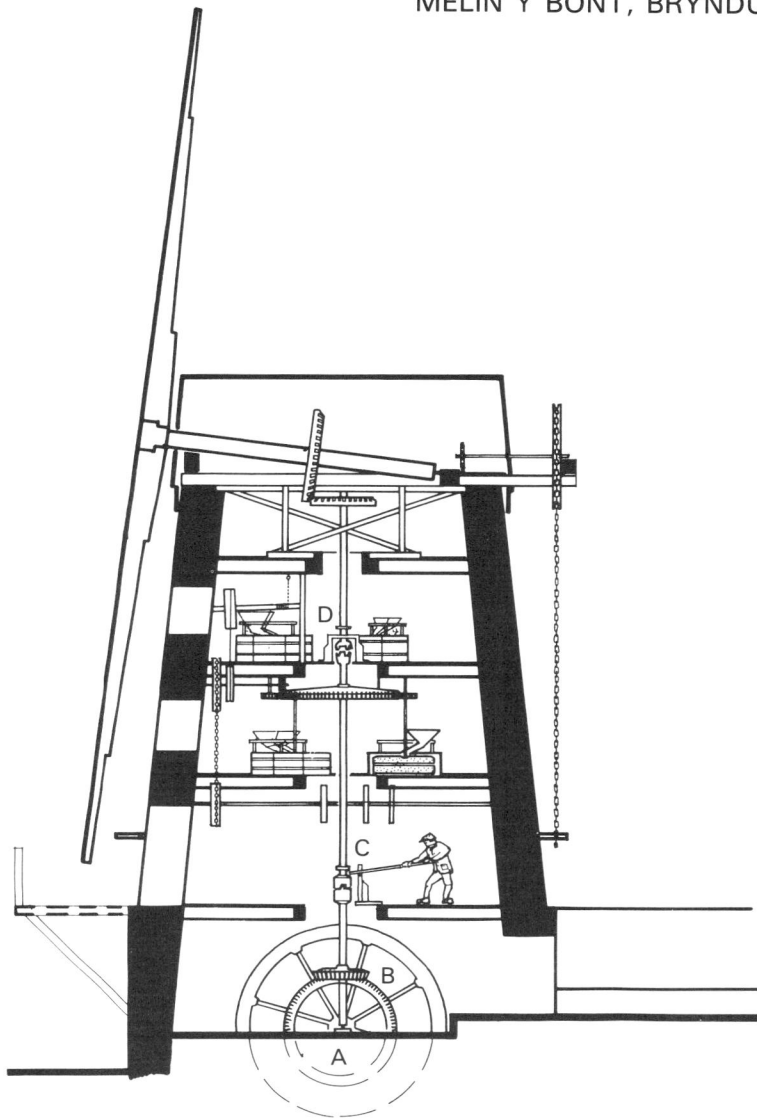

Drawn by Gerallt Nash and reproduced by permission of the Welsh Folk Museum.

Plate 50 Melin y Bont reflected in its mill pond c. 1913. Melin Maelgwyn can be seen on higher ground in the distance.

Plate 51 The mill's cast-iron waterwheel had wooden paddles 4' 6'' in length. A sluice gate, operated by rack and pinion gearing, controlled the flow of water to the wheel.

Plate 52 John Williams, the last miller to work Melin y Bont, photographed in 1929 lifting the drive from the waterwheel out of engagement.

Plate 53 A toll board in the mill displays the charges for grinding wheat, barley and oats.

for a straw chaffer, swede cutter and other machines.

Sadly, the ivy-clad mill now stands derelict, gutted by fire in April 1973 when two small children playing with matches accidentally set light to hay being stored inside. Weakened by the flames the wooden floors gave way depositing the mill's contents in a charred heap at the bottom of the tower, only the cast-iron ring bearing for centring the cap remaining in place. On its way down the iron upright shaft snapped in two as it smashed into the waterwheel, fracturing part of the wheel's outer rim in the process.

The broken machinery still lies there today, gradually being submerged under a deepening layer of pigeon droppings. Outside, the mill stream has reverted to its original course and the pond, once filled with trout, is now dry, a grassy depression where silage is stored and farm equipment parked.

LLANFA!R MATHAFARN EITHAF

Melin Rhos Fawr
Brynteg

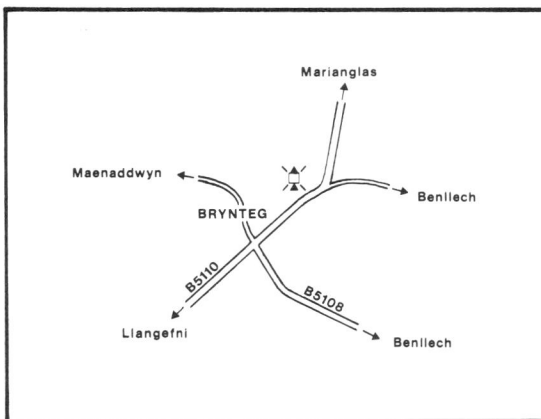

SH 497829

The empty shell of Melin Rhos Fawr stands about half a mile north-east of Brynteg crossroads, at the side of the B5110 to Marianglas.

Built in 1757 it appears to have had an uneventful working life until becoming immortalised as 'Mona Mill' in George Borrow's book **Wild Wales** (1862). In his search for places connected with the celebrated Anglesey poet Goronwy Owen, Borrow had met John Jones, 'Melinydd of Llanfair', and been invited to his house for refreshment. John Jones was tenant of Melin Rhos Fawr, 'a man about thirty, rather tall than otherwise, with a very prepossessing countenance'. After a meal of bread and cheese and sugary tea, Borrow accompanied John Jones to the mill 'which lay some way down a declivity, towards the sea'. Near the mill was a 'comfortable- looking house' (still standing today) which belonged to the owner of Melin Rhos Fawr. On meeting him in the mill yard Borrow was greeted in Welsh and asked if he had come to buy hogs!

John Jones remained at Melin Rhos Fawr until his death in 1877, aged 57. His widow Martha took over the tenancy and is named as miller in **Slater's Directory** of 1883 and again as such in **Sutton's Directory** of 1889. She died in 1889. The mill continued in operation for a few more years, run by William Jones of Brynteg, before finally closing around 1910. The cap and sails were then taken down and the machinery sold for scrap.

In 1978 planning permission was granted to convert Melin Rhos Fawr into a dwelling, but work was never started and the permission has now lapsed. The tower, though roofless, is still in reasonable condition and is used as a store.

Plate 54
Melin Rhos Fawr in 1903. The barn has been replaced by a bungalow and the cornfield is now pasture.

LLANFIHANGEL YSGEIFIOG

Melin Berw
Pentre Berw

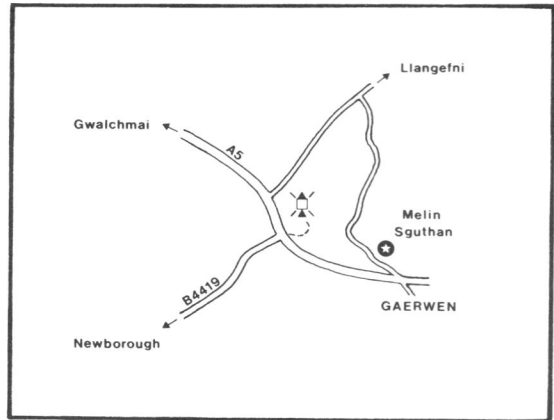

SH 474723

Melin Berw is the most westerly of three windmill towers lining the north side of the A5 between Pentre Berw and Gaerwen. Access to it is via a steep uphill track opposite the junction of the A5 and the B4419 to Newborough.

Built on the edge of a rocky bluff, Melin Berw was ideally situated to take advantage of the westerly winds sweeping in across the flat expanse of Malltraeth Marsh. The mill probably dates from the late eighteenth century and,

Plate 55
Melin Berw soon after it ceased to be used as a dwelling.

TO BE SOLD BY

AUCTION,

BY MR. W. DEW,

AT THE

BULL'S HEAD INN,

IN THE TOWN OF LLANGEFNI,

On Thursday, the 8th day of August, 1850,

Subject to conditions to be then and there produced, unless disposed of in the meantime by Private Contract, of which due Notice will be given,—ALL THOSE

TWO WATER CORN CRIST MILLS,

AND ALL THAT

WIND CORN GRIST MILL,

CALLED

BERW MILLS,

TOGETHER WITH THE

MACHINERY, DRYING-KILNS, DWELLING-HOUSES, GARDENS, STABLES,

AND APPURTENANCES THERETO BELONGING,

AND ABOUT 5 ACRES OF EXCELLENT LAND

SITUATE IN THE

PARISH OF LLANIDAN,

IN THE COUNTY OF ANGLESEY,

AND NOW IN THE OCCUPATION OF WILLIAM ROBERTS AND ELLEN ROBERTS.

The above Premises are held under a Lease for the life of a person aged 50 years, and 14 concurrent years, to be computed from the 13th day of November, 1849, at the reserved Rent of £9 per annum.

The Mills and Buildings are in thorough repair and most advantageously situated contiguous to the Post Road leading from Menai Bridge to Holyhead and in the immediate vicinity of the Berw Collieries, within the distance of three miles of the market-town of Llangefni, and one mile of the Gaerwen Station on the Chester and Holyhead Railway.

William Roberts, the Tenant, will shew the premises and further particulars may be obtained on application to

F. JONES, PRINTER, BEAUMARIS.

MR. O. OWEN, SOLICITOR, BEAUMARIS.

together with two nearby watermills, formed part of a complex called Berw Mills. In 1850 all three mills were put up for auction, the sale notice particularly stressing their advantageous location in relation to road and rail links.

However, this favourable location did not prevent Melin Berw from becoming the first of the three windmills in the vicinity to cease working. Precisely when, or why, this occurred is not known but soon after its closure sometime towards the end of the nineteenth century the mill was converted into a residence, complete with new roof and brick chimneys. This change of use may explain why the 1901 six-inch O.S. map made no mention of a disused windmill even though a circular building was marked. If this was an oversight on the part of the cartographer it was rectified on the 1926 edition, the words 'Old Windmill' appearing, although according to Rex Wailes' survey three years later Melin Berw was still being used as a dwelling.

By the early 1930s Melin Berw was no longer occupied and the building had begun to deteriorate. Before long roof, chimneys and most of the top floor had gone. In recent years attempts have been made to arrest this decline and the abbreviated tower now possesses a corrugated plastic roof which provides shelter for poultry being reared inside. Lately, a breeze-block cattle byre has been added to one side of the tower.

Plate 56
Miller and family outside Melin Maengwyn c.1890.

Gazetteer

LLANFIHANGEL YSGEIFIOG

Melin Maengwyn
Gaerwen

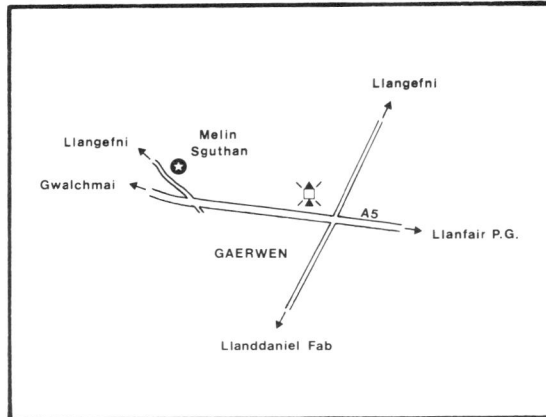

SH 485720

Three quarters of a mile east of Melin Berw is the empty shell of Melin Maengwyn. It stands on a grassy mound behind a row of houses which border the A5 next to Gaerwen's 'Little Chef' restaurant.

The mill is early nineteenth century in origin, a stone tablet set in rubble masonry above its north-facing doorway being inscribed with the date 1802 and the letters H^{W}_{E}. The initials are those of H.E. Williams, and successive generations of the Williams family appear to have run Melin Maengwyn throughout its working life, although they did not own it. Originally part of the Plas Newydd Estate, the mill and surrounding farmland were bought sometime around 1860 by Hugh Pritchard, a successful baker from Liverpool. His descendants now farm at Tygwyn, Llanddaniel Fab and still retain ownership of Melin Maengwyn.

According to **Slater's Directory**, the miller in 1883 was William Williams of Gaerwen, but whether the apparently uninterested man in the photograph, taken towards the end of the century, is William Williams or his successor R.J. Williams is not known for certain. The last miller to work Melin Maengwyn was Hugh Williams who shrewdly married into the Pritchard family, thus uniting landlord and tenant. For the last few years of its life the mill operated with only two sails, relying on a steam traction engine to provide additional power. Its career was finally brought to a close by a severe storm just after the First World War; lightning struck the endless chain, fusing the metal, and strong winds then toppled the cap.

Today the roofless tower serves as a shelter for cattle and survives in reasonable condition, despite a gaping hole in the stonework.

LLANFIHANGEL YSGEIFIOG

Melin Sguthan
(Union Mill)
Gaerwen

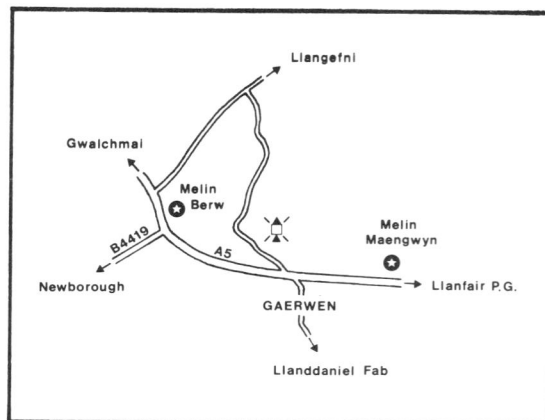

SH 478722

Midway between Melin Maengwyn and Melin Berw is the third of the windmills in the locality. Known variously as Melin Sguthan, Union Mill or Gaerwen Mill, the capless tower stands just off the A5 at the side of the minor road to Craig Fawr and Llangefni.

The use of the name Gaerwen Mill on early O.S. maps and estate documents would seem to indicate that this mill predates Melin Maengwyn, the village's other windmill, and is therefore late eighteenth century in origin. Unfortunately, Gaerwen Mill's early history, like that of its neighbour, is largely unrecorded. At some time during the second half of last century, however, ownership appears to have passed into the hands of an association of Manchester people — hence the alternative name Union Mill.

Thomas Parry, the present occupant of Union Mill cottage, believes the union in question had some connection with canals and remembers a union badge, the design of which incorporated a sailing ship, hanging for many years above a door in the adjoining stables. This lends support to the link with Manchester as the city's coat of arms contains such a ship and can, with permission, be used by any Manchester organisation including local unions or craft guilds. Nonetheless this still begs the question why such a union should be interested in owning a windmill on Anglesey. Although 'anti-mills' were established last century on co-operative principles in response to the monopoly held by local landowners there is no evidence for this being the case in Gaerwen.

During the latter part of the nineteenth century Gaerwen Mill was worked by Rowland Williams. In the photograph he stands, in best suit and bowler hat, next to the building in which were stabled the two horses and carts used for local collections and deliveries. The external stairway provided access to an upper floor which functioned as a kiln for roasting oats.

Like most millers Rowland Williams kept pigs and the low building in front of the windmill served as a pigsty (long since demolished). Pigs were of par-

Plate 57
Gaerwen Mill c.1890. Unlike the majority of Anglesey's windmills Gaerwen Mill employed a 'Y-wheel' type of chain wheel for turning the cap. This had iron forks like Y-shaped pegs at intervals round the wheel's circumference which were thought to give the endless chain a better grip.

ticular importance to a miller because they could be fattened on the grain taken in lieu of monetary payment and sold to provide a cash income. This system of deducting a percentage of a farmer's grain as payment, however, was open to abuse by unscrupulous individuals and contributed to the miller's reputation (often undeserved) for not being entirely honest. On Anglesey an oft-repeated saying among farmers went as follows :

> 'The miller's pigs are always fat,
> But whose grain are they eating?'

According to Thomas Parry Gaerwen Mill ceased working in 1913 and four years later was deliberately set on fire to facilitate the recovery of its metalwork for the war effort. Unfortunately, the fire also seriously weakened the structure and several bad cracks are now apparent, visual evidence of its dangerous condition.

Plate 58
Two boys pose on a broken sail of Melin Wynt y Craig shortly after its closure.

Gazetteer

LLANGEFNI

Melin Wynt y Craig

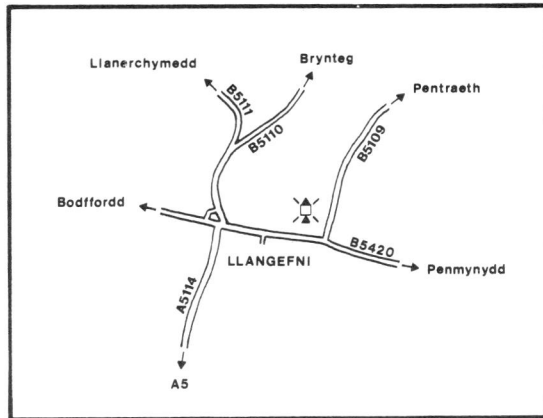

SH 465757

Perched on top of a rocky eminence, the ruined tower of Melin Wynt y Craig dominates the eastern skyline above the busy market town of Llangefni. It stands beside the B5109 to Pentraeth, about 200 yards from its junction with the B5420.

Although ideally situated to receive the wind from all directions, Melin Wynt y Craig must have been awkward for local farmers to reach with their sacks of grain. A steep crag face blocks the approach on its western side, while elsewhere access is by narrow, rocky paths.

The mill is believed to have been built in 1829. *Pigot's National Commercial Directory* of 1828 records only a watermill in Llangefni, but the windmill must have come into existence soon afterwards because in the 1833 edition of the same directory William Hughes is named as miller of the Craig Mill. By 1849 William Hughes had retired and Craig Fawr Mill, as it was called in *Slater's Directory* of that year, was being run by Roland Owen. Whether he was the Owen of 'Williams & Owen' mentioned in the same directory as being in charge of Llangefni's watermill is not known. This partnership seems to have ended in the early 1850s because only Thomas Williams is recorded at the watermill in 1858. He was succeeded by William Williams, possibly a relation of Robert Williams who was working the windmill around the same time.

According to R.O. Roberts, William Jones was the miller of Melin Wynt y Craig when it closed in 1893. He also presided over the last years of two other mills and 'for that reason was nicknamed ''Angau Melinau'', a sort of human death watch beetle, who brought all these mills to a standstill.'

By the turn of the century Melin Wynt y Craig had begun to fall into disrepair and soon afterwards the remains of its sails were taken down and its machinery removed. Today, the empty shell with its blocked up windows and wooden flagpole functions only as a landmark, clearly visible for miles.

Plate 59
Owen Hughes and his dog pose for the camera in front of Melin Llangoed.

LLANGOED

Melin Llangoed
(Tros y Marian)

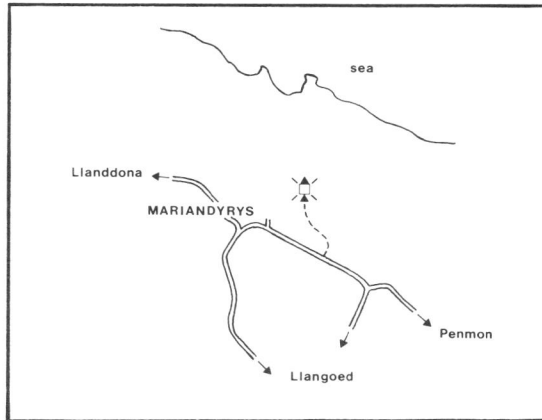

SH 608812

Melin Llangoed stands on a hilltop about one mile north of Llangoed village and occupies one of the most impressive sites on Anglesey. The view from the windmill tower encompasses much of southern Anglesey, Snowdonia, the Menai Strait, Puffin Island (Ynys Seiriol) and the Great Orme.

This exposed position no doubt influenced Henry Williams' choice when deciding on a suitable location for the erection of his windmill in 1741. Constructed from locally quarried limestone Melin Llangoed was one of the earliest of Anglesey's tower mills, only Melin Gallt y Benddu (Llanerchymedd) known definitely to predate it.

The mill remained in the ownership of Henry Williams until 1787 when it was sold, together with adjacent farmland, to John Hughes of Caernarvon. An extract from the Articles of Agreement between the two parties describes the sale in the legal language of the time: 'Henry James Williams doth . . . promise and agree to . . . grant convey and assure unto the said John Hughes . . . all that capital Messuage or Tenement with the Desmesne and other lands, Hereditaments and Premises with the Appurtanances thereto belonging . . . And all that Wind Mill, with other Appurtanances to the same belonging commonly called or known by the Name of Trosymarian Wind Mill. The said John Hughes . . . shall . . . well and truly pay to the said Henry James Williams . . . the full sum of One thousand, six hundred and fifty five pounds of lawful money of Great Britain for the absolute purchase of the same Premises.'

Information concerning the mill over the next fifty years is scant, but it must have changed hands at least once, being advertised for auction in 1842 as 'a good strong-built Wind Corn Grist Mill situate on Tros-y-Marian Estate . . . in the occupation of the Rev William John Lewis.'

Melin Llangoed eventually became the property of a prosperous local landowner called Major Chadwick who rented out the mill, its cottage and accompanying farmland. In ***Slater's Directory*** of 1883 the miller is named as Owen

Jones. A seaman by profession Owen Jones spent the first part of his working life travelling the world on the sailing ships of the day. On retirement from his maritime career he took over the tenancy of Melin Llangoed and worked the mill until its closure in 1921.

In 1926 all the metalwork except the cast-iron windshaft was removed for scrap and the mill remained an empty shell until 1960 when it was bought by Stanley Flory, a civil engineer. Concerned about the lack of an efficient water supply to his house near the mill, Flory put his professional knowledge (gained as construction engineer at the Shanghai Waterworks) to good use by installing a water tank in the hollow tower. He diverted water from a nearby spring through sand filters to a reservoir in the field below his house (Tan- y-Felin). From here it was pumped uphill into the tank in the mill, thus providing a reliable supply of fresh water for both his house and the cottage adjacent to the mill. During conversion, the mill's old windshaft was utilised to help support an internal floor, and a spiral stairway was fitted to give access to a viewing platform at the top of the tower.

Plate 60
Melin Cefn Coch.

LLANFECHELL

Melin Cefn Coch

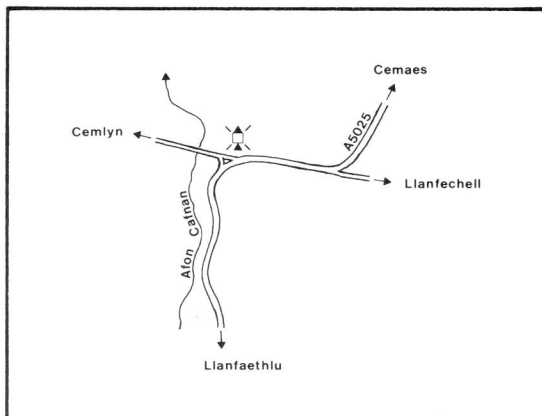

SH 342914

The truncated tower of this small mill can be seen from the A5025 midway between Llanrhyddlad and Tregele, near its junction with the minor road to Cemlyn Bay.

Although only part of the original structure still stands, that which remains is in relatively good condition, being constructed from narrow slabs of a grey-green Precambrian rock common across northern Anglesey. Both doorways and the surviving window have cambered lintels of rough-cut vertical slabs, giving the tower a touch of individuality. The interior is devoid of machinery and open to the elements, while outside broken pieces of Welsh stone lie scattered about, half concealed by vegetation.

Melin Cefn Coch is known to have been working in the late eighteenth century, but precisely when it was built is uncertain. William Bulkeley, who lived only two miles away, makes no mention of it in his diaries and as these cover most of the years between 1734 and 1760 it seems probably that the mill originates from the 1760s or 1770s.

Few facts have come to light concerning its working life, and even its date of closure remains unclear. The mill is not shown as disused on the 1901 six-inch O.S. map but this could represent an error on the part of the surveyor, for when the nearby house 'Tyn-y-Felin' appeared for sale in the same year Melin Cefn Coch was described as 'Old Windmill'. According to the mill's owner who farms at Caerdegog Uchaf the tower has been in its present state for as long as anyone can remember. This may explain why it was overlooked by Rex Wailes when he carried out his survey of the island's windmills in 1929.

Interestingly, Melin Cefn Coch was only one of a number of different mills operating in the locality. The others, all watermills, harnessed the energy of the Afon Cafnan, a small river issuing from Llyn Llygeirian and flowing northwards to the sea at Porth-y-Pistyll. Because it powered an unusually large number of mills along its two mile course the Cafnan became known locally as 'Afon-

115

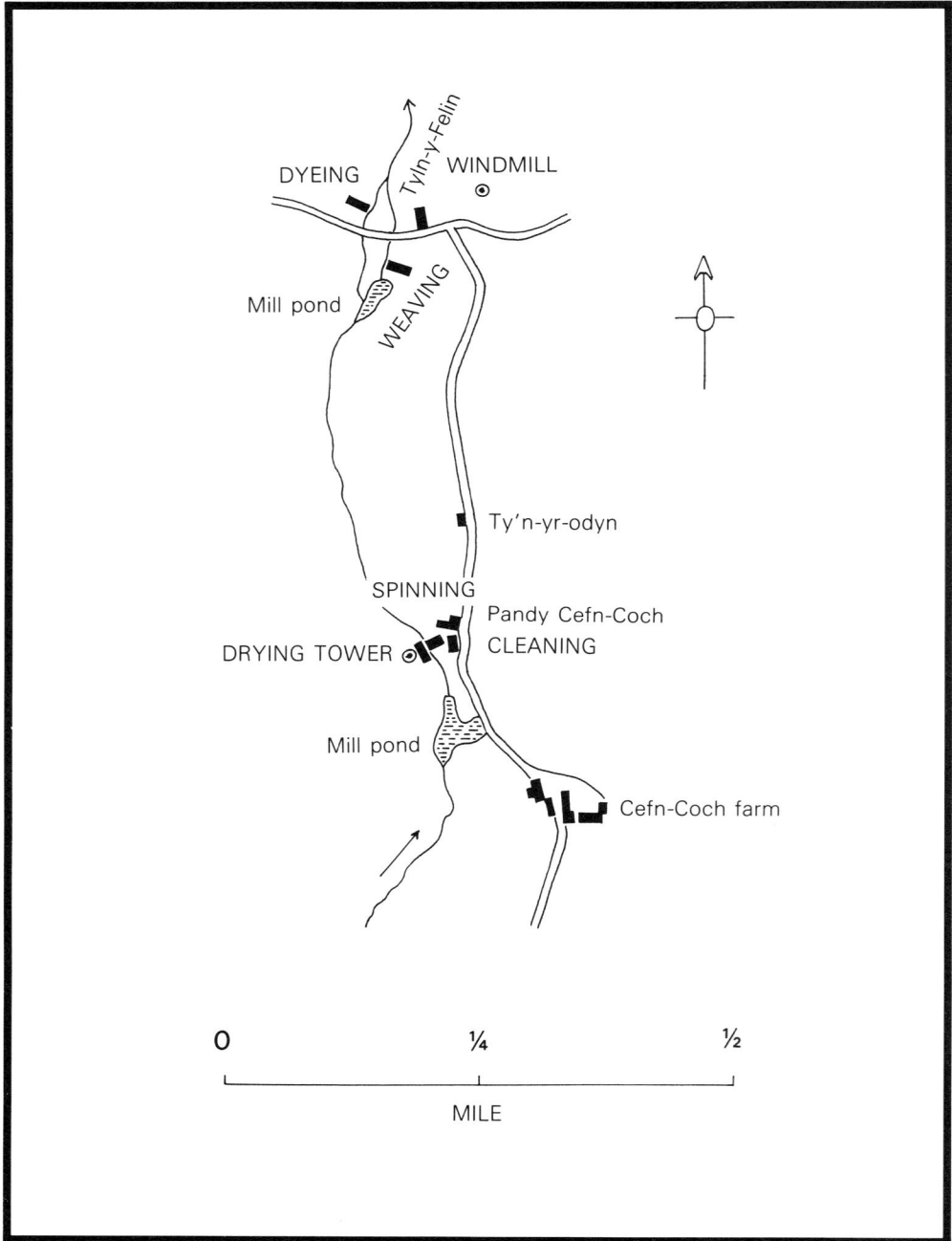

DYEING

Ty'n-y-Felin

WINDMILL

Mill pond

WEAVING

Ty'n-yr-odyn

SPINNING

Pandy Cefn-Coch

DRYING TOWER

CLEANING

Mill pond

Cefn-Coch farm

0 ¼ ½

MILE

PANDY CEFN-COCH WOOLLEN COMPLEX

y-Pandy', Pandy being Welsh for fulling mill, a building in which woollen cloth was cleaned. In such a mill the cloth was immersed in a trough containing a mixture of urine, soda, fuller's earth and water and then pounded by wooden hammers driven by a waterwheel.

On Anglesey the spinning and weaving of woollen cloth were essentially domestic occupations carried on in farmhouse or cottage, usually by women. Fulling was the only part of the cloth making process which required a separate building and weavers came to depend on their local Pandy to finish the material. At Cefn Coch the initial function of the Pandy was supplemented by the introduction of machinery for carding, spinning, weaving and dyeing — resulting in a small industrial complex along the banks of the Afon Cafnan.

The river supplied a millpond at Pandy Cefn Coch where fleeces were washed before being hung to dry in a windmill-like drying tower. Spinning the woollen fibres into thread took place in an adjoining building, its machinery powered by a waterwheel fed from the millpond. A short distance downstream the spun thread was woven into cloth at another mill, while a separate leat from its pond carried water to a third mill in which the woven cloth was dyed. Thus, within a third of a mile stretch of the Afon Cafnan the whole process of cloth making took place, from fleece to finished material.

This small-scale production of woollen cloth, mainly to meet local needs, was common practice throughout Anglesey well into the nineteenth century. However, as communications with the mainland improved the island's protective isolation was gradually removed, exposing rural industrial concerns like Pandy Cefn Coch to competition from the much cheaper products of larger, purpose-built textile mills.

Pandy Cefn Coch managed to struggle on until the end of last century before finally closing. When the Cefn Coch Estate was broken up for sale in 1901 the description of Lot 32 seemed poignantly appropriate: 'Cefn Coch Woollen Factory and Mill at one time carrying on an extensive business but of late operations have been suspended'.

As well as powering the woollen mills discussed above, the Afon Cafnan also supported two corn mills. One has long since disappeared, but Melin Cafnan (SH 345933), now owned by the National Trust, still retains much of its original machinery.

Plate 61
Melin Mechell from the south-east. A footpath next to the wall leads down to a derelict watermill
which was once worked in conjunction with the windmill.

MECHELL

Melin Mechell
(Melin Minffordd)

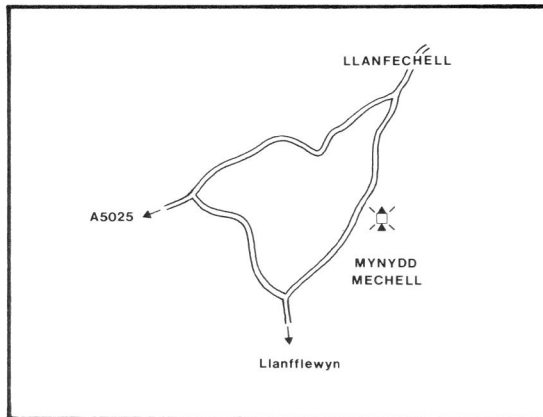

SH 362902

At the northern edge of the scattered settlement of Mynydd Mechell, approximately midway between Llanfechell and Llanfflewyn, is the converted tower of Melin Mechell. With its replica wooden cap and slate-roofed, single storey extension, Melin Mechell provides one of the better examples of a windmill conversion on Anglesey, successfully managing to maintain the character of the original building.

The date of the mill's construction remains unknown, but by the latter part of the nineteenth century it was being run by the Parry family of Nuadd Filwa, Llanfechell. Robert Parry is named as miller in **Slater's Directory** of 1883 and also in **Sutton's Directory** six years later. By 1889 Morgan Parry, presumably Robert's son, had taken over the mill and probably worked it until its closure around the beginning of the First World War.

When Rex Wailes surveyed Melin Mechell in 1929 the cap had gone and only one sail survived, hanging from the rusting windshaft. Surprisingly, most of the internal machinery was still in place, one unusual item being a double flour dresser comprising two cylinders within a single casing.

The mill continued to deteriorate until the late 1970s when planning permission was obtained for its conversion into a dwelling. On completion of the work Melin Mechell was advertised for sale as a "carefully restored windmill, providing excellent 3-bedroomed accommodation, on ⅛ acre site". In 1982 the mill was purchased by Mr and Mrs Kevin Flannery. On moving in they discovered several pieces of old mill machinery which they donated to the millwrights who were then restoring Melin Llynon, Llanddeusant.

Plate 62
Melin Pant y Gwydd is now part of a house.

MECHELL

Melin Pant y Gwydd

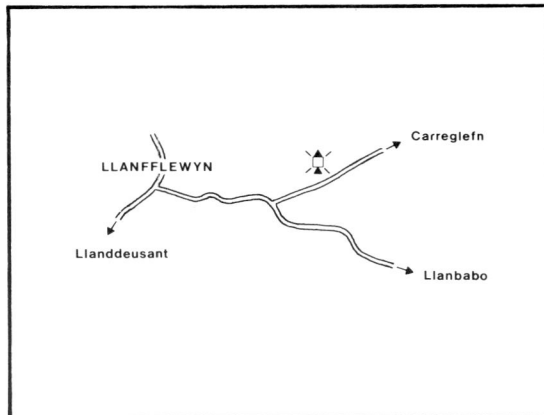

SH 365887

From Melin Mechell a minor road winds south and then east towards the village of Carreglefn. After one and a quarter miles a narrow, gorse-lined track leaves the road and climbs to a whitewashed house attached to which is the tower of Melin Pant y Gwydd.

Virtually nothing is known of the history of this remote mill. It is thought to date from the first half of the eighteenth century and to have stopped working early in the nineteenth, but documentary evidence to confirm this is slight. The 1844 Tithe Map reveals that the property (consisting of six acres and two roods) was occupied by John Edwards, who paid 14*s*.7*d*. rent in lieu of tithes to Hugh Davies Griffith. However, there is no reference to a windmill on the site which suggests that Melin Pant y Gwydd had closed by then, a theory supported by the absence of a windmill symbol on the first edition of the one-inch O.S. map of 1840.

Although Melin Pant y Gwydd occupies an excellent position for a windmill, open to the wind from all directions, it is surrounded by extremely poor farmland, unsuitable even for oats. Gorse-topped, rocky outcrops cover extensive tracts in the immediate vicinity, interrupted only by an occasional patch of level ground. It is therefore likely that grain for Melin Pant y Gwydd had to be transported fairly long distances and this may have contributed to its early closure.

The tower, which has lost some of its original height, now forms the central part of a private dwelling, its circular shape almost obscured by later additions.

Plate 63
Little now remains of the tower of Melin Orsedd.

PENTRAETH

Melin Orsedd
Rhoscefnhir

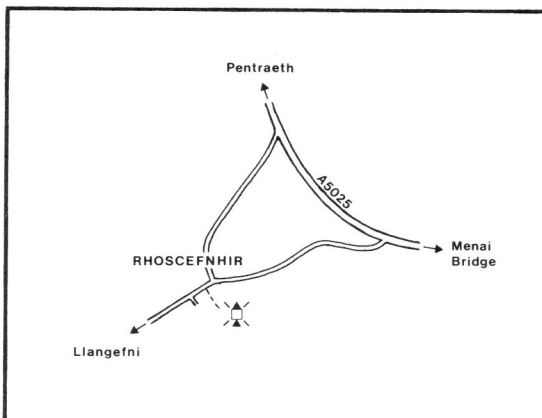

SH 523763

Not much is left of Melin Orsedd. The windmill, which once dominated the small hamlet of Rhoscefnhir just over a mile south of Pentraeth, is now little more than a stump. A narrow opening between two cottages gives access to the truncated tower which stands on a grassy mound overlooking the flat, marshy valley of the Afon Braint.

The semi-circular remains vary in height from around twelve feet on the north-west side to less than four feet on the south-east. Two doorways with cambered gritstone lintels face each other across blocks of stone which have fallen from the crumbling wall. A tenacious ash sapling pushes out of this rubble towards the light.

The date of Melin Orsedd's construction is unknown, and its early millers unrecorded. In *Slater's Directory* of 1856 John Jones is named as miller of Melin Orsedd, but who followed him is not clear, the 1883 and 1889 editions of the Directory listing both Thomas Jones (possibly John Jones' son) and Robert Williams. Perhaps they ran the mill jointly, although only the latter's name appears on an invoice for meal dated March 1888. It is likely that either, or both, were the last to work Melin Orsedd as the mill is believed to have burned down in the early years of this century. As far as elderly residents of Rhoscefnhir can remember the present structure has changed little in their lifetime.

Plate 64
Melin y Gof in full sail being admired by two visitors to the mill. Its last set of sails was made by John Roberts of Melin Cemaes and his brother Owen of Llangefni, the canvas coming from a Holyhead sailmaker called Ellis.

One of the miller's tasks was to dress the stones when their cutting edges wore down. Alfred Jones remembered this having to be done quite frequently at Melin y Gof due to the carelessness of farmers who left bolts and other pieces of metal in amongst the oats! The special tools he used for this work were made by local blacksmith John Williams.

TREARDDUR

Melin y Gof
(Stanley Mill)
Trearddur Bay

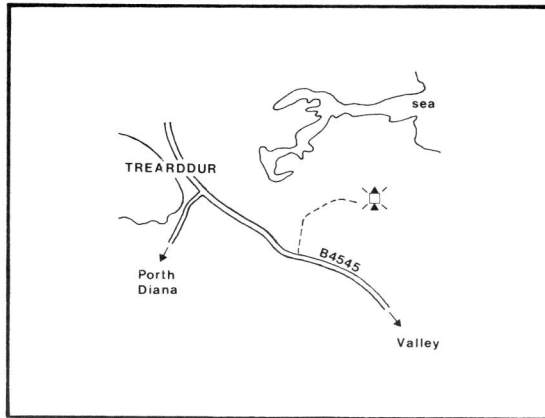

SH 266788

Located approximately three quarters of a mile east of the holiday resort of Trearddur Bay, Melin y Gof occupies an elevated position overlooking the Inland Sea. It is reached by a long track which leaves the B4545 at the southern edge of the village.

Construction of Melin y Gof is thought to have begun in 1826 (if a date scratched in the stonework is to be believed) and been completed a year or so later. When Rex Wailes visited the mill in 1929 he discovered a runner stone still in use bearing the date 1828, probably the year in which Melin y Gof came into operation. Considering its location, surprisingly little has come to light of the mill's early history, other than its date of construction and that it was built on land owned by the Stanley family of Penrhos.

By the 1880s Melin y Gof was being run by Robert Morris who lived at 'Ynys y Gof', the farm next to the mill. He was followed towards the end of the century by Owen Owens, who later moved to Rhoscolyn. Alfred Jones then took over Melin y Gof and was still there when it was badly damaged by a storm in November 1938. For some reason confusion seems to have arisen over the precise date that the mill closed. The years 1934 and 1936 have both been quoted in print, *The Times* newspaper even publishing a half page photograph of Melin y Gof in its edition of 16th September 1936 with the caption: 'it has just ceased working and was one of the last of the windmills in Anglesey to be in use.'

However, according to Alfred Jones, interviewed a few years before his death in 1987, both he and his brother Hughie had been milling the evening prior to the fateful storm. Worn out they had retired to bed and soon fallen asleep. During the night the wind had increased in strength, changing direction to behind the sails. The partially rotten cap had offered little resistance and both cap and sails had come crashing to the ground.

The following account was written soon after the event by Lucy Williams,

Plate 65
Melin y Gof photographed in 1939, a few months after its destruction by winter gales.
''In the midst of the wreckage could be seen the great cast-iron windshaft, snapped clean in two with the shock of the impact. The proud sails were reduced to splintered chaos, one still arm pointing skywards, and the cap had disappeared, carried away by the wind.''
'Anglesey and the North Wales Coast' by F.H. Glazebrook.

Melin y Gof today.

later to became an authority on Anglesey's tide mills. 'During the recent gales Holyhead has lost one of its most beautiful and striking landmarks, the old wind-mill near Trearddur Bay. This mill was the last to work in North Wales . . . grinding supplies of oats, barley and Indian corn for neighbouring farmers. There are still people who remember having wheat ground there for household use, and the exquisite nut-like flavour of that kind of Anglesey bread is among one's pleasant recollections. This mill was stripped of its sails, cap and outside gear by the recent storm and a mere melancholy tower remains instead of the mov-ing beauty of the windmill.'

Although a scheme to repair and preserve Melin y Gof was initiated, the Second World War intervened and people's attention became focussed on other, more pressing matters. With no protective cap the mill's vulnerable interior was exposed to the wind and rain, hastening its decay. Neglected during the war years and ignored subsequently, Melin y Gof continued to deteriorate, nobody even bothering to remove the broken sails which lay next to the tower.

In the early 1960s planning permission was obtained to turn the now derelict mill into a dwelling. During the conversion work the height of the tower was raised by one storey to provide a studio room with eight windows which af-forded extensive views of the surrounding countryside.

In the past, being able to see neighbouring mills at work was very helpful to a miller. By observing them he could glean information about changes in the strength and direction of the wind and adjust his sails accordingly. Alfred Jones recalled the day when three of his friends arrived at Melin y Gof. Visibili-ty was good and four other windmills could be observed in the distance, prompting the trio to compose the following rhyme, each person contributing two lines. Translated from the Welsh it reads . . .

'Machreth mill is grinding,
Llynon mill is working.
Caergeiliog mill has stopped dead,
Ynys y Gof is going ahead.
I can see another mill,
Melin yr Ogof is going still.'

Not exactly eisteddfod material, but nonetheless a salutory reminder that wind-mills were once sufficiently numerous for several to be seen from a single van-tage point.

Plate 66
Built into the side of a long barn, Melin Geirn now stands empty. The holes for the support timbers of its gallery can be seen in what remains of the protective render.

TREF ALAW

Melin Geirn

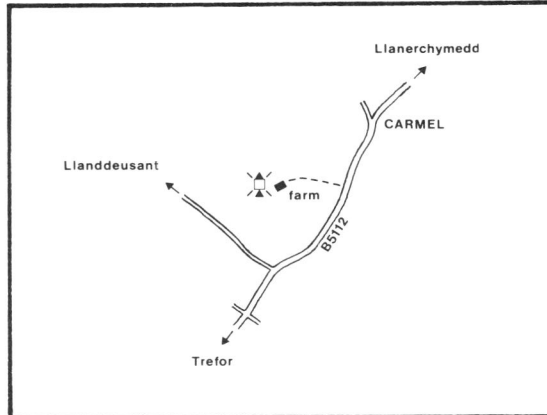

SH 3828 18

Melin Geirn is part of a farm of the same name situated just off the B5112, half a mile south-west of the hamlet of Carmel.

Already out of use by the end of the nineteenth century, Melin Geirn was the tallest of a small number of domestic windmills which were attached to, or integrated with, other farm buildings. Over the years the five storey tower has lost most of its protective rendering, revealing it to be rather crudely constructed from rough rubble and the minimum of mortar. A doorway at second floor level once gave access to a gallery from which the sails could be reached for adjustment. The height of the gallery suggests that the sails were shorter than usual, perhaps in order to clear the roof of the adjoining barn.

The ruined farm windmill of Treban-Meirig prior to demolition.

FFERAM.

Is composed of superior Agricultural Land in good state of cultivation, with excellent Fences, Gates, &c.; is also approached and intersected by capital Roads. The situation commands pleasant views of the surrounding country, and is within easy distance of the villages of Pencaerneisiog, Gwalchmai and Aberffraw, and also of Bodorgan and Ty Croes Stations on the Chester and Holyhead Railway.

THE AGRICULTURAL BUILDINGS

Are modern and arranged with every convenience, and comprise as follows:—

Boiling House, Range of Piggeries, Poultry House, Cattle Sheds, Cow Houses, Calf House, Root House, Large Barn with Apparatus for grinding corn, and Granary over; Men's Lofft, Hay and Straw Rooms, Chaff House, Cart Horse Stable, fitted with water connection, Horse Box, Nag Stable, &c., with a large enclosed Yard.

Coach House, Cart House, WINDMILL with all necessary machinery and connections for Churning; Chaff Cutting, Pulping, Gorse Cutting, &c., (now in thorough working order); Rick Yard, &c., &c.

Possession to be had on the 13th day of November, 1873.

In other examples it is likely that the proximity of adjacent buildings would have reduced the efficiency of the farm windmill by restricting the full rotation of the cap and sails. On smaller mills of this type, such as Treban- Meirig and Fferam near Llanfaelog (both now demolished), the cap was either fixed or able to turn only through a limited angle.

Treban-Meirig was built against the gable end of an eighteenth century barn and contained a single pair of millstones. Fferam, on the other hand, probably had no stones at all. When the farm and agricultural buildings appeared for sale in 1873 no mention was made of the tower containing millstones for grinding corn. However, it did possess 'all necessary machinery and connections', indicating that some sort of drive was taken directly from the windmill into the adjoining barn to power various items of farm machinery including, as the sale notice points out, the 'Apparatus for grinding corn'.

Farm windmills, therefore, tended to be multi-purpose and grinding corn was just one of the many tasks they performed.

The ruined farm windmill of Fferam prior to demolition.

Plate 67
Melin Llynon in 1929, starting to show signs of neglect.

TREF ALAW

Melin Llynon
Llanddeusant

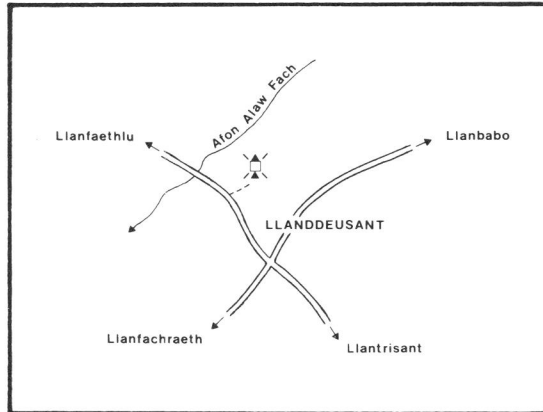

SH 341852

Melin Llynon lies at the centre of a tangle of narrow lanes between Llan-fachraeth and Llyn Alaw, half a mile north-west of the crossroads at Llanddeu-sant and not far from Melin Howell, Anglesey's only water-powered corn mill still in operation.

Melin Llynon was built by Andrew Williams for surgeon Herbert Jones of Llynon Hall whose solicitor, Hugh Ellis, preserved several documents relating to the mill's construction. These reveal that it cost £529.11s. and took seven months to build, from August 1775 to March 1776, during which time Andrew Williams appears to have gone bankrupt. Nevertheless, Melin Llynon was com-pleted and Thomas Jones became its first miller. When he died, aged 90, in 1846 Melin Llynon passed to his son of the same name who worked it for 23 years until his death in 1869. Although his two sons, Hugh and Henry, then inherited the mill they were not enthusiastic about running it and the 1881 Cen-sus shows that responsibility for Melin Llynon had been transferred to William Pritchard their cousin.

Over a century's connection with Melin Llynon ended for the Jones family in 1892 when Robert Rowlands, another of the seven milling brothers, took over the mill and adjoining farm. He enjoyed a long and successful partnership with Melin Llynon, managing even to see it through the difficult years of the First World War when strict Government regulations controlled the production of flour. However, in August 1918 disaster struck when a severe storm damaged the cap so badly that it became permanently locked in one position. Afterwards very little grinding took place, the mill being able to operate only when the wind blew from the south-west. Despite Rowlands' offer of payment towards the cost of repairs Melin Llynon's owner refused to sanction the work, regretfully conceding that the mill's useful life had come to an end.

Following the closure of the mill Robert Rowlands stayed on at the mill house, continuing to farm the land and run his flour distribution business. His large, horse-drawn wagon would collect flour from ships unloading at Valley Cob and return to the farm where customers' orders would be made up and delivered in a cart driven by his carrier William Williams.

Robert Rowlands appears to have been a well-liked and much respected man, known locally as a gifted musician. His granddaughter, Mrs. Eirwen Williams, spent her childhood at Melin Llynon and remembers him devoting much of his free time to running the children's choir. He also allowed the mill to be used for rehearsals by the parish male voice choir and the local drama group. Mrs. Williams moved into the village with her grandfather when he retired in 1929 and looked after him until his death ten years later. He was buried next to his second wife, Ellen, in the churchyard at Rhydwyn, the village of his birth.

Plate 68
Robert Rowlands photographed in 1933 aged 84.

TREWALCHMAI

Melin Gwalchmai

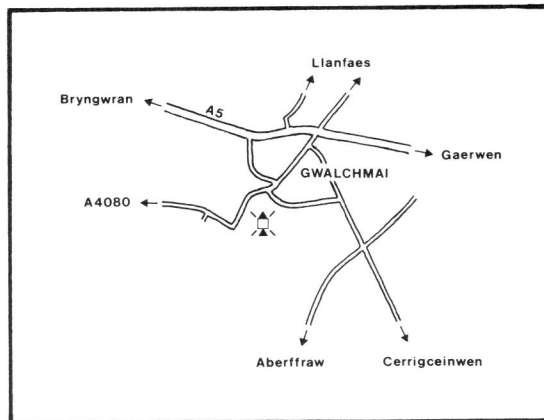

SH 385759

Melin Gwalchmai is situated just over quarter of a mile from the centre of Gwalchmai village, by the side of a steep, narrow lane which winds south-west to meet the minor road to Pencarnisiog. The mill's elevated position makes it clearly visible from the A5, so it is puzzling how Rex Wailes came to overlook Melin Gwalchmai when compiling his list of Anglesey's windmills in 1929.

Melin Gwalchmai probably dates from the early nineteenth century, although little is known of its history except that it was built on land belonging to the Treveilyr Estate. From at least the middle of last century it was looked after by successive generations of the Williams family. William Williams appears as miller in **Slater's Directory** of 1849 and in later editions up to 1883. By 1889, according to **Sutton's Directory**, the mill had passed to Morris Williams, but his tenure must have been relatively brief because six years later Robert Williams is named as miller. He continued to run Melin Gwalchmai until well into the twentieth century before handing over to his nephew Thomas Williams. The mill was still being worked by wind as late as 1927, but not long afterwards the sails were taken down and the boat-shaped cap replaced by a flat, concrete roof.

Grinding continued for several more years using a diesel engine which Thomas Williams had installed in a corrugated iron shed next to the mill. The drive to the millstones passed underground into the mill's cellar, a room entirely enclosed within the mound on which the tower stands. The tower itself is still in very good condition and although all the machinery has gone the wooden floors and steps survive, albeit somewhat worm-eaten. Remarkably, one Welsh bedstone remains in place on the stone floor. Pieces of the other millstones, including several segments of French burrstone, can be seen outside incorporated in a boundary wall.

Plate 69
Robert Williams, his family and carrier pictured outside Melin Gwalchmai on a postcard sent in January 1908. The entrance to the mill's cellar is below the figure furthest right.

INDUSTRIAL WINDMILLS

Parys Mountain Windmill

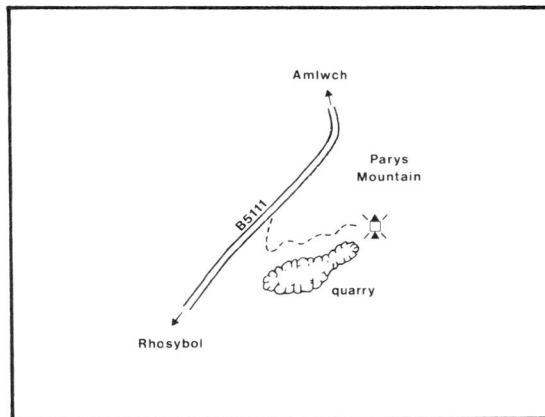

SH 444906

Not all Anglesey's windmills were constructed to grind corn or drive farm machinery. On the summit of Parys Mountain, one and a half miles south of Amlwch, is the empty shell of a tower mill built to pump water from the underlying copper workings. Today the ruined tower presides over a desolate and sterile landscape dominated by a huge open-cast pit created when Parys Mountain was the world's largest copper mine.

The remarkable rise of the island's copper industry followed the discovery in 1768 of a rich vein of ore known as the Great Lode. Within a few years the mines were employing 1,500 people and producing 80,000 tons of copper annually. However, the boom was short-lived and by the beginning of the nineteenth century decline had set in. Although fortunes revived briefly in the early 1830s the abolition of the duty on imported copper ore meant that further exploitation became uneconomic. Mining still continued, but by the end of the century less than 500 tons of ore were being raised annually.

It was during this period, in 1878, that the mining company, in an attempt to reduce operating costs, decided to erect a windmill on the top of Parys Mountain. Its purpose was to assist an existing steam engine to pump out water, raise ore and transport workers in the nearby Cairn's Shaft. The windmill's five sails (unique on Anglesey) drove bevelled gearing which connected via an upright shaft to a crankpin at the base of the tower. This transmitted reciprocating motion through 200 feet of long wooden rods to the crankpin of the steam engine. Thus, on windy days, the use of the windmill resulted in a considerable saving of coal all of which had to be carted up to the summit boiler. This arrangement appears to have been relatively successful, at least until the beginning of the twentieth century; in a mining treatise published in 1901 the author commented that 'the windmill, constructed by Captain Hughes, is working well'.

Flooding had been a problem since mining began. In the early years water was raised in buckets attached to *whimseys* — primitive windlasses fixed along

Plate 70
Parys Mountain pumping windmill c.1910. The remains of the connecting rods to the steam engine can be seen bottom left.

Plate 71
Parys Copper Mine, Anglesey by J. Warwick Smith (1785)

138

the upper edge of the open-cast pit. Whimseys were also employed to lower men to the workings and to bring the copper ore to the surface. Most whimseys relied on men or horses for their operation, and although it seems reasonable to assume that windpower was also used the evidence for this is not conclusive. A letter from Thomas Williams to the Earl of Uxbridge in 1788 listing the purchase of various items of equipment for the mines did mention ''a windmill engine'', but unfortunately included no details of its intended use.

The whimseys depicted in a watercolour painting of the mineworkings dated 1785 by J. Warwick Smith (1749-1831) are all hand or horse powered. Although a small, straight-sided windmill can be observed in the distance it is difficult to be certain whether it was a mine pump or the corn mill which, according to a map of 1785, stood on the side of Parys Mountain near Cerrig y Bleiddiau farm. Interestingly, in another painting of the same area executed five years later the windmill has gone, presumably a casualty of the expanding open-cast workings. If it was the corn mill its loss was probably not greatly mourned at the time, the higher wages offered by the copper mines having lured local people away from farm work and led to a general neglect of agriculture. Also, the toxic pollution emanating from the burning of copper ore in open kilns to remove the sulphur content had destroyed the vegetation in the vicinity and with it the livelihood of farms such as Cerrig y Bleiddiau.

Even today little grows in this blighted landscape, only a few patches of heather managing to survive amid the bare rock and multi-coloured spoil heaps.

St Eilian Colour Works Windmill

SH 449913

Not far from the copper mines and close to the trackway which carried copper ore to the smelters of Amlwch stood the St. Eilian Colour Works. Here a windmill was used to crush yellow ochre and other coloured earths for the manufacture of paint.

Given the wealth of documentary evidence which exists about the copper mines, surprisingly little is known of the origins and activities of the adjacent Colour Works. However, the business must have been well-established by 1850 for, according to **Slater's Directory** of that year: 'The most prominent branch [of industry in Amlwch] besides mining is the manufacture of colours and paints.' Under the heading 'Colour Manufacturers and Merchants (Ochre, Venetian Red, Umber, etc.)' the Directory gives the name Richard Parry, Saint Ellan Mills.

Ochre was a by-product of the process by which copper was recovered from the mineral-rich water issuing out of the mineworkings. This water was channelled into long, shallow pits filled with pieces of scrap iron. Over time some of the iron dissolved, to be replaced by a coating of copper oxide which could then be scraped off. Heavily impregnated with iron sulphate the water was allowed to slowly run off and, exposed to the air, the iron sulphate precipitated as hydrated iron oxide or yellow ochre. After collection and drying the ochre was

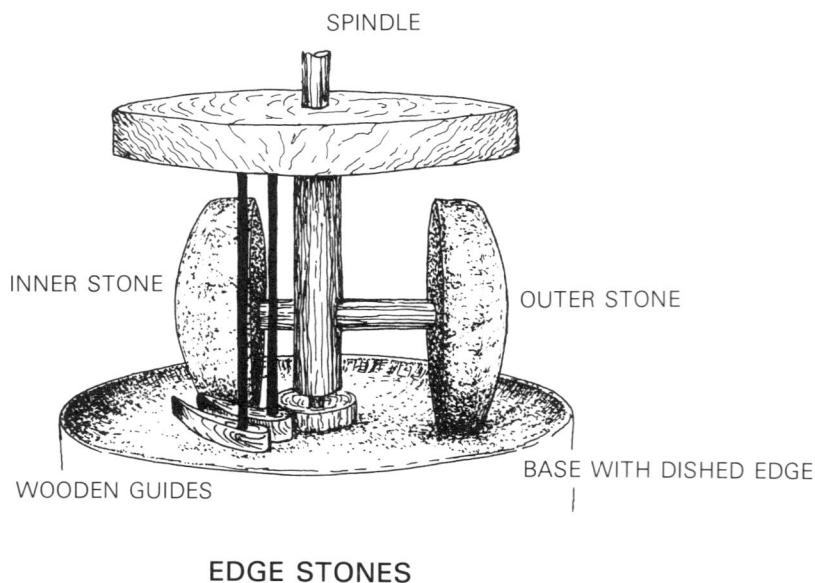

SPINDLE

INNER STONE   OUTER STONE

WOODEN GUIDES   BASE WITH DISHED EDGE

EDGE STONES

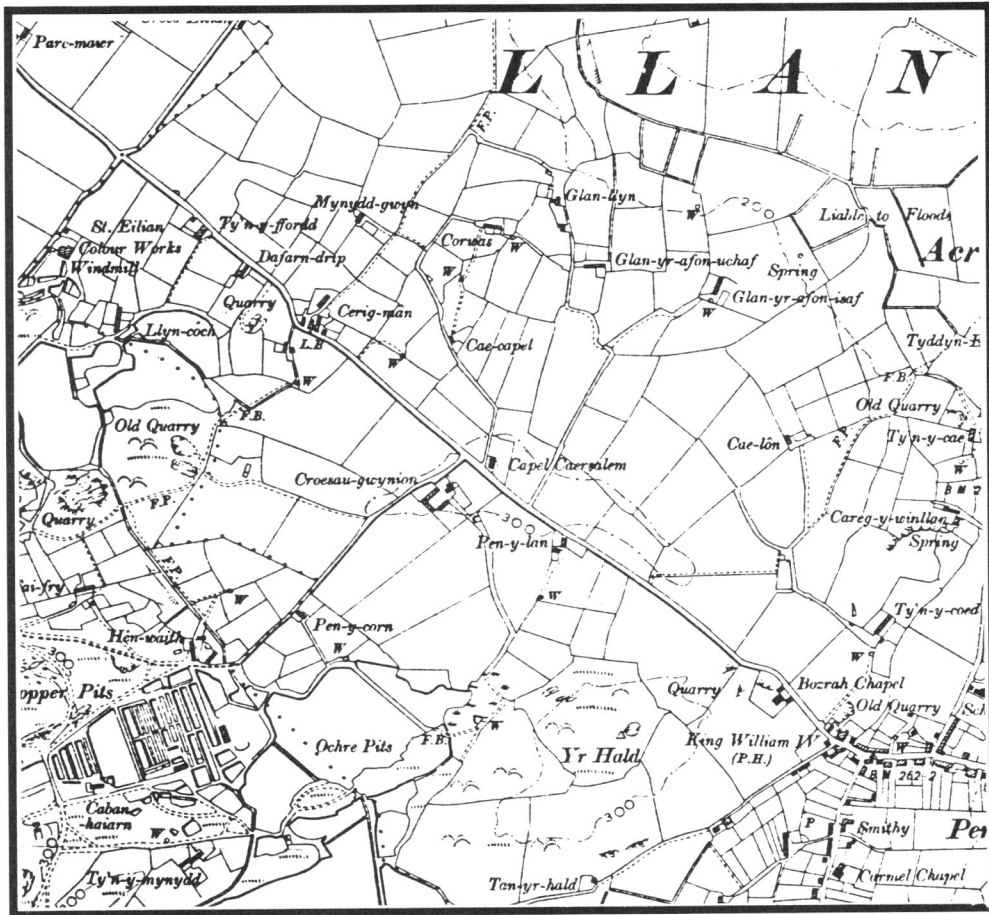

Site of the now demolished St. Eilian Colour Works windmill. Reproduced from the 1901 Ordnance Survey map.

ready for crushing in the windmill. The millstones used for this purpose differed in several respects from those of a traditional corn mill. Three stones found on the site of the now demolished Colour Works proved to be relatively small (26 inches in diameter) but quite thick (12 inches), with rounded edges and slightly bevelled sides. Their faces were smooth, no dressings being necessary as the edges of the stones did the grinding. Inside the mill two upright stones were fixed on a horizontal axle which passed through a vertical spindle and operated rather like a cider press. One stone was set nearer the central spindle than the other in order to roll a different path over the crushing pan. Ochre was fed under the stones by wooden guides and when reduced to powder was swept off the pan into a receptacle below.

Plate 72
The old tower of Tan y refail windmill shortly before its removal.

Tan y Refail Windmill, Holyhead

SH 248816

A third windmill employed for non-agricultural purposes stood in Kingsland, Holyhead, on a rocky crag above the Tan y Refail yard of William Williams the building contractor (now the premises of R.J. Roberts Commercials).

The windmill supplied power to the sawmill and joinery works below by means of a connecting drive shaft which passed from the bottom of the mill tower into the roof of the workshop (which was at the same level). The shaft ran the whole length of the building and belt drives were taken from it to operate the saws. For a number of years the windmill worked in conjunction with a steam mill, and both can be seen in the illustration on the firm's billhead, dated 1914. However, a later billhead points to the discontinuation of wind power because, although the same picture had been retained, the word 'Wind' no longer appeared in the title. The disused tower survived until 1971 when it was demolished to enable the building of an access road to two new bungalows overlooking the yard.

William Williams, who died in 1899, built many of the chapels in Holyhead and was for several years superintendent in the construction of the town's famous breakwater. His firm also owned a cargo steamer called the 'Hercules' which plied regularly between Holyhead and Liverpool from a small dock in Newry Beach Yard.

	COMMUNITY	WINDMILL
1	Amlwch	Melin Adda
2	Amlwch	Melin y Borth
3	Amlwch	Melin y Pant
4	Bodffordd	Melin Frogwy
5	Bodffordd	Melin Manaw
6	Bodffordd	Melin Newydd
7	Bodorgan	Melin Hermon
8	Cwm Cadnant	Melin Llandegfan
9	Cylch y Garn	Melin Drylliau
10	Holyhead	Melin yr Ogof
11	Llanbadrig	Melin Cemaes
12	Llanddyfnan	Melin Llanddyfnan
13	Llanddyfnan	Melin Llidiart
14	Llanerchymedd	Melin Gallt y Benddu
15	Llanfaelog	Melin Maelgwyn
16	Llanfaelog	Melin y Bont
17	Llanfair-Mathafarn-Eithaf	Melin Rhos Fawr
18	Llanfihangel Ysgeifiog	Melin Berw
19	Llanfihangel Ysgeifiog	Melin Maengwyn
20	Llanfihangel Ysgeifiog	Melin Sguthan
21	Llangefni	Melin Wynt y Craig
22	Llangoed	Melin Llangoed
23	Mechell	Melin Cefn Coch
24	Mechell	Melin Mechell
25	Mechell	Melin Pant y Gwydd
26	Pentraeth	Melin Orsedd
27	Trearddur	Melin y Gof
28	Tref Alaw	Melin Geirn
29	Tref Alaw	Melin Llynon
30	Trewalchmai	Melin Gwalchmai
31	Amlwch	Parys Mountain Windpump

THE KNOWN LOCATIONS OF DEMOLISHED

TOWER MILLS

a Aberffraw
    Fferam (SH 361736)

b Amlwch
    St. Eilian Colour Works (SH 449913)

c Bryngwran
    Treban Meirig (SH 367771)

d Cylch y Garn
    Rhydwyn (SH 315889)

e Holyhead
    Llanfawr (SH 256817)

f Holyhead
    Tanyrefail (SH 249815)

g Holyhead
    Ucheldre (SH 243823)

h Llanddona
    Llanddona (SH 578798)

i Llaneilian
    Llanwenllwyfo (SH 477895)

j Llanerchymedd
    Rhodogeidio (SH 410851)

k Llanfachraeth
    Llanfachraeth (SH 312829)

l Llanfair yn Neubwll
    Caergeiliog (SH 307784)

m Llangristiolus
    Capel Mawr (SH 413717)

n Llangristiolus
    Rhostrehwfa (SH 439748)

o Llanidan
    Brynsiencyn (SH 482672)

p Mechell
    Felin Nant (SH 393900)

q Rhosyr
    Dwyran (SH 445654)

Plate 73
Wind turbines of the type proposed for Parys Mountain.

Postscript

Anglesey has a long history of wind power dating back to the Middle Ages, and although it is unlikely to witness a resurgence of the corn grinding wind-mill, there is a possibility that in the near future a new generation of wind machines will be seen in the landscape.

Using the wind to generate electricity is not a new idea; small machines have provided power for lighting in remote areas of Britain since the middle of this century. However, in recent years, attention has been focussed on renewable sources of energy because of fears for dwindling fossil fuel reserves and concern about the environmental damage they cause when burnt. Wind power is regarded as one of the more promising alternative sources for development, due in part to considerable advances in the technology involved.

So far, the contribution made by wind to Britain's electricity supply has been small, but plans for large wind farms are under consideration and pioneering projects such as the one at Carmarthen Bay in south-west Wales already exist. In the last year a number of firms have shown interest in Anglesey as a possible location for the generation of electricity utilising wind power. Among the schemes is a proposal by Anglesey Mining plc to erect eight 400 kW turbines on Parys Mountain.

The high initial cost of installation of the turbines and the current prices paid for electricity would normally make it impossible for such a wind farm to be economically viable. However, the Government has recently introduced a Non-Fossil Fuel Obligation which requires the electricity companies to purchase, at a higher price, a quota of electricity not generated from burning fossil fuels, at least until 1998. The belief that this stipulation will allow wind energy to compete favourably with fossil fuels in Britain has encouraged MANWEB and Vestas (the Danish manufacturer of wind turbines) to join with Anglesey Mining plc in this venture.

The wind turbines proposed for the Parys Mountain site are of the same design which Vestas has already successfully erected in Denmark and elsewhere. Each turbine consists of three 37' (17.4m) blades mounted on a self- supporting tubular steel tower 105' (32m) in height, fixed to concrete foundations. If planning permission is granted the wind farm could be generating electricity by summer 1992. The company aims to produce eight million kilowatt hours of power per year which will replace part of the conventionally generated electricity presently used in the mining operations.

Not surprisingly, concern has been expressed about the impact of such a scheme on the environment — the towers themselves would far exceed the present height limit of structures on Anglesey. However, Parys Mountain is already an industrial site and there are no dwellings in the immediate vicinity of the proposed wind farm. Also, there would be no interference with radio or TV equipment, nor any increase in traffic once construction was complete. Anglesey Mining plc consider the visual impact of the current mining operation would possibly be improved by the wind farm, which would be blended into the existing environment without any detriment to the surroundings.

Nevertheless, the question remains as to whether people will find the wind turbines environmentally acceptable, even though this obviously ''clean'' form of energy will help combat the problems of harmful emissions and waste from conventional and nuclear power stations. Will the tall, slim towers with their slowly revolving, aerodynamically efficient blades be seen as modern incarnations of the picturesque windmills once common on Anglesey, or will they be regarded in the same light as the intrusive electricity pylons which march across the countryside? Only time will tell.

WHERE IS IT?

148

Cymdeithas Melinau Cymru

Welsh Mills Society

The Welsh Mills Society was launched in October 1984. The aims of the Society, as stated in its Constitution, are '' . . . to study, record, interpret and publicise the wind and water mills of Wales; to encourage general interest; to advise on their preservation and use, and to encourage working millers . . . '' (A copy of the Society's Constitution will be forwarded upon request).

The Society's Annual General Meeting is held in October and a Spring regional meeting, alternating annually between north and south Wales, is held in late April. The Welsh Mills Society arranges lectures, films, visits to mills, demonstrations and courses on mill recording techniques and millwrighting as well as actively publicising mills which are open to the public. Regular newsletters are sent to all members, together with the Society's annual journal, MELIN, containing articles on subjects relating to the Welsh mills scene.

For further details please contact the membership secretary: Miss J. F. Roberts, ''Cerrig Melin'', 39 Keene Avenue, Rogerstone, Newport, Gwent NP1 9DF.

Plate 74
The photograph is from the Isaac Hughes collection of Michael and Judy Kerr of Llanberis and was taken around 1885. Isaac Hughes was an itinerant photographer who travelled throughout North Wales in the latter part of the nineteenth century and left behind a large collection of glass plate negatives. Although the windmill is typical of those found on Anglesey it has, so far, eluded identification. If any reader recognises the mill, either from the building itself or from the people standing outside it, the authors would be pleased to hear from them. Any other information relating to Anglesey's windmills, especially old photographs or postcards, would also be gratefully received.

Bibliography

Windmills

Beedell, Suzanne	Windmills	*David & Charles, 1975*
Brown, R. J.	Windmills of England	*Robert Hale, 1976*
De Little, R. J.	The Windmill, Yesterday and Today	*John Baker, 1972*
Freese, Stanley	Windmills and Millwrighting	*Cambridge University Press, 1957; reprinted by David & Charles, 1971*
Shillingford, A. E. P.	England's Vanishing Windmills	*Godfrey Cave Associates, 1979*
Vince, John	Discovering Windmills	*Shire Publications, 1968*
Vince, John	Power Before Steam	*John Murray, 1985*
Wailes, Rex	The English Windmill	*Routledge & Kegan Paul, 1954*
Wailes, Rex	Windmills in England	*Architectural Press, 1948; reprinted by Charles Skilton, 1973*

Anglesey

Carr, A.D.	Medieval Anglesey	*Anglesey Antiquarian Society, 1982*
Eley, Geoffrey	Mona, Enchanted Island	*The Priory Press, 1968*
Ramage, Helen	Portraits of an Island: Eighteenth Century Anglesey	*Anglesey Antiquarian Society, 1987*
Richards, Melville (ed)	An Atlas of Anglesey	*Anglesey Community Council, 1972*
Senior, Michael	The Island's Story	*Gwasg Carreg Gwalch, 1987*
Williams, E. A.	The Day Before Yesterday: Anglesey in the Nineteenth Century	*G. Wynne Griffiths, 1988*

The Transactions of the Anglesey Antiquarian Society also provide a valuable source of reference, and contain the only previous study of the island's wind and watermills, viz.

The Mills of Anglesey by R. O. Roberts (1958)

Index

Entries in *italics* refer to publications mentioned in the text. Page numbers in bold type refer to page illustrations.

154

Snowdonia 113
Society for the Protection of Ancient Buildings 37
Somerscales of Keelby 53
Stanley family 35, 71, 79, 125
Stanley Mill (see Melin y Gof)
Stead, F. 43
steam power 13, 33, 35, 84, 85, 107, 137, 143
stock 15
stone dressing 19, **19**, 124
stone nut 16, 17, **81**
Suthkroc (see Abermenai)
Sutton's Directory 60, 71, 87, 103, 119, 135

Tal y Bont 6
Tan y Refail Windmill 58, **142**, 143
Telford, T. 1
tenancy 5, 31, 73, 79
tentering 20
tentering screw **20**
The Cambrian Traveller 1
The Miller 62
The Times 125
Thompson and Son, R. 36, 37, 43, 45, **52**, 53, 55
tide mills 6, 7, 73, 127
Trearddur Bay 6, 11, 15, 25, 35, 37, 79, 125
Trearddur Bay Mill (see Melin y Gof)
Trefor 69, 70
Tregele 115
Tre'rgo 6
Treveilyr Estate 135
Tros y Marian (see Melin Llangoed)
Tunnicliffe, C.F. 41, 43
twist peg 17
Twrgarw 6
Ty Croes 97
Tyddyn Olifer (see Melin Hermon)

Union Mill (see Melin Sguthan)

Valley 33, 35, 134
vat 21

Wailes, R. 36, 37, 41, 43, 45, 55, 76, 85, 88, 106, 115, 119, 125, 135
wallower 17
Warwick-Smith, J. **138**, 139
water power 2, 5, 6, 8, 31, 38, 39, 60, 66, **67**, 68, 71, 73, 74, 90, 97, 98, **100**, 105, 111, 115–117, 133
water tower 76, 114
weather **14**, 15
weather conditions 2–3, 8–10, 90
weather damage 8, 13, 25, 33, 41, 43, 71, 85, 93, 96, 107, 125, 127, 133
Welsh Church Acts Committee 35, 36
Welsh Mills Society 38, 86
Welsh stone 17, 21, 115, 135
West Mill (see Melin Ucheldre)